Unsettled Waters

CRITICAL ENVIRONMENTS: NATURE, SCIENCE, AND POLITICS

Edited by Julie Guthman, Jake Kosek, and Rebecca Lave

The Critical Environments series publishes books that explore the political forms of life and the ecologies that emerge from histories of capitalism, militarism, racism, colonialism, and more.

Unsettled Waters

Rights, Law, and Identity in the American West

———

Eric P. Perramond

UNIVERSITY OF CALIFORNIA PRESS

University of California Press, one of the most distinguished university presses in the United States, enriches lives around the world by advancing scholarship in the humanities, social sciences, and natural sciences. Its activities are supported by the UC Press Foundation and by philanthropic contributions from individuals and institutions. For more information, visit www.ucpress.edu.

University of California Press
Oakland, California

Cataloging-in-Publication data is on file with the Library of Congress

ISBN 978-0-520-29935-1 (cloth)
ISBN 978-0-520-29936-8 (paperback)
ISBN 978-0-520-97112-7 (e-edition)

27   26   25   24   23   22   21   20   19   18
10   9   8   7   6   5   4   3   2   1

CONTENTS

ILLUSTRATIONS, MAPS, AND TABLES

ILLUSTRATIONS

MAPS

TABLES

PREFACE

"Why are you here, *exactly*?" Tamara asked me on a hot July day in 2011. We stood by the Rio Santa Barbara, a high-mountain New Mexican stream lined with cottonwoods and aspens, with dry meadows and juniper-dotted hills stretching beyond. Tamara had been describing her work as a *mayordoma*, the water ditch boss in charge of allocating the stream's trickle to her neighbors' fields.

By that point, four years into my research, I'd been asked this question many times, and I had the answer: "I'm here to listen and learn."

"Well," she said, straightening up a bit, "that's different."

Unsettled Waters conveys the voices and concerns of actual people caught up in the legal labyrinth of what are known as *water rights adjudications*. Water adjudications in the American West are often state-driven lawsuits designed to find, map, and inventory existing water-use rights. Every western state has its own distinct procedure, although many share similar templates.

For all involved in these watershed lawsuits—the irrigators, lawyers, technicians, politicians, and observers like me—adjudication is a complex, adversarial, and sometimes baffling process. Adjudication has been underway for well over a century, longer than New Mexico has been a state, and there is no end in sight. It can also seem boring, as most of the work is conveyed in legalese through administrative court documents. This is partially why the process has often been ignored.

However, as I learned from water users like Tamara, adjudication is anything but boring. Adjudication resurfaces conflicts. It delves into the intricacies of state cartographies, of multiple histories of colonialism, culture, sovereignty, and identity. The process has exposed, antagonized, and rearticulated social relationships over water that are vital to understand in an era of increasing water scarcity and competing demands.

This book is a geographic ethnography that makes use of living testimony, historical and legal archives, and on-the-ground observations from New Mexico. When I moved back to the West in 2005, I was interested in pursuing regional research that reflected my new academic appointment at Colorado College in Southwest studies and environmental science. It all started as a modest project to document how small ditches in New Mexico were coping with water scarcity. It turned into something far more complex than I had anticipated. Adjudication kept coming up in my interviews. I was curious as to what adjudications entailed, why they were taking so long, and what they meant for water users. My small summer project turned into a decade-long study that, like adjudication, still continues. Like the process itself, this book has had to account for the state's fixation with tracking water users and prior dates of water use and the fact that the process is ongoing and still happening in the present, with an eye toward the future. After all, when New Mexico started this enterprise, the hope was that there might be water left over to allocate.

As New Mexicans made clear from the start, our collective water futures are at stake in adjudications. I listened to water experts across the spectrum, from irrigators like Tamara to the attorneys and state experts who conduct the process. *Unsettled Waters* is a book for all who care about the future of water, the ways in which states allocate and manage water, and the effects of these largely unseen legal proceedings on water users.

ACKNOWLEDGMENTS

I am most grateful to all the water rights holders, irrigators, ditch bosses (mayordomos), ditch riders, water managers, lawyers, engineers, hydrologists, and personnel from the Office of the State Engineer (New Mexico) who generously agreed to share information, insights, and expertise of all kinds. This book is the result of our conversations, and I hope it provokes more of them.

Various chapters and snippets in *Unsettled Waters* were informed by Melanie Stansbury, David Correia, Juan Estevan Arellano, Darcy Bushnell, Eric Shultz, Sylvia Rodriguez, Stanley Crawford, Miguel Santistevan, Maria Lane, David Benavides, Paul Mathews, David Garcia, and Aaron Bobrow-Strain. I want to single out Maria Lane's excellent work that is reshaping our understanding of water, science, and the courts during the territorial period of New Mexico. Melanie Stansbury and Darcy Bushnell were vital to my early understanding of the Aamodt case and the multiple outcomes of adjudication in general. Long-time water authors Helen Ingram and Jim Wescoat corrected early misconceptions and filled in gaps. New Mexico water beat reporters Staci Matlock and John Fleck also had useful feedback and insight as I developed the prospectus for this book. Rick Carpenter at the City of Santa Fe Water Division also provided repeated individual and course visits to the infrastructure of that small city.

To the late Juan Estevan Arellano, I owe the most: his lessons and his memory remind us of the values and challenges of small-scale irrigation in New Mexico. His recently released *Enduring Acequias* appeared before he passed in late October 2014 and will no doubt serve his legacy well. William Doolittle and the late Karl Butzer shaped my early understanding of the value of ancient and historic irrigation systems.

At Colorado College, I am humbled to have current (and former) colleagues willing to share their time, support, and expertise. This work was shaped by conversations, ideas, and inspiration by my colleagues in Southwest studies—notably, Santiago Guerra and Karen Roybal (Montoya). Their scholarship and teaching remind me that the greater Southwest is a source of theory and praxis, not just a recipient. Colleagues Anne Hyde, Takeshi Ito, David and Christina Torres-Rouff, and Corina McKendry shared their early and late version insights on this project. Informal discussions over beer were no less vital, and I thank William Davis, Michael O'Riley, and Tyler Cornelius for often listening to my book-driven rants. I have benefited from students' comments, insights, and shared senior capstone moments, as well as their own senior-year work that informed parts of *Unsettled Waters*.

The Colorado College Crown Faculty Center (Rebecca Tucker, Jane Murphy) supported a development workshop on the first iteration of this manuscript, during which James F. Brooks, Stanley Crawford, and Wendy Jepson shared comments and insights that vastly improved the organization, tone, and argumentation you find here. The Office of the Dean, Social Science Executive Committee (SSEC), and multiple Jackson Fellowships from the Hulbert Center for Southwest Studies funded this work between 2007 and 2016. Additional funding from the SSEC in 2017 paid for Bill Nelson's finely executed maps. A sabbatical in 2017 and 2018 provided the time to write this book.

An initial conversation with Rebecca Lave led me to the Critical Environments book series at the University of California Press and Kate Marshall and Bradley Depew, who provided guidance throughout. Peer reviews at the University of California Press by Tom Sheridan and Maria Lane provided valuable and constructive criticism. The book production process was ably assisted by Nicholle Robertson at BookComp, Inc., and I thank the excellent work of copy editor Wendy Lawrence. I am grateful to artist Chuck Forsman for allowing me to use his painting *Native Land* as inspiration for the book cover and the Colorado Springs Fine Arts Center at Colorado College for permission to use this work.

Archival collections and their excellent staff were fundamental: the New Mexico State Records Center and Archives, the Center for the Southwest at the UNM main library, the small collections in the basement of the Office of the State Engineer's Bataan building complex, and the Fray Angélico Chávez Historical Library were all carefully consulted. I benefited from being an ACM (Associated Colleges of the Midwest) Newberry Library Faculty fellow during the fall semesters of 2013 and 2017, in Chicago. At Newberry, Diane Dillon, Scott Stevens, Jim Akerman, and 2013–2014 Newberry fellows Tobias Higbie, Kathleen Washburn, Michael Vorenburg, Leon Fink, Elizabeth Shermer, Susan Sleeper-Smith, Patricia Marroquin Norby, and Michael Schermer were especially helpful in shaping an initial prospectus for this book. Diana and J. Stege allowed us time for writing in their idyllic adobe house in New Mexico, and they have my profound thanks.

The personal debts here run deep. I thank my close and extended family for letting me work and live like a hermit this past year. To Ann, my constant companion and compassionate, ruthless critic, I owe you most of all.

Introduction

The Cultures of Water Sovereignty in New Mexico

I walk with Hector along the irrigation canal, the village's *acequia,* as the ditch is called in northern New Mexico. The only sounds are the pulsing, burbling water in the constrained channel and fluttering cottonwood leaves. Standing on the canal bank, we can feel the vibrations from the water through our feet. Hector, the mayordomo, a kind of ditch boss, turns and raises his eyebrows: "It's pretty clean, isn't it?"

"The water looks great," I respond.

He frowns, shaking his head, and starts walking again as he mutters, "No, I mean the banks of the acequia."

I stop. In one awkward moment, I missed the point completely. From Hector's perspective, this is not about the water. His remark is about the collective work of villagers in nature. The acequia is not *just* a ditch; it is also an important institution in which people manage and allocate water.[1] Water brings together the people on the ditch, through the act of sharing it, and it is the lifeblood of this valley. The ditch and the institution are *their* work, defining *their* water and their landscape. As a cultural and political institution, the acequia keeps neighbors from fighting over water. These ditch institutions have forged an agrarian cultural landscape with political clout and clear institutional rules. After this momentary misunderstanding, I forge on.

"So, Hector," I ask, as he walks along his ditch, "are your water rights adjudicated and accounted for on this stream?"

He turns, cocking his head, a slight squint in one eye, coming to a dead stop. "Who wants to know?"

I am sure I blink a few times. After a beat, I respond. "I do."

His head lobs farther toward his shoulder. The squint hardens. "And who are you, exactly? You're not a lawyer are you, or some fancy engineer?"

I'm not wearing a suit, but my words suggest I have one hidden under my jeans and T-shirt. I rush to explain I am neither a lawyer nor an engineer. I'm a geographer from a liberal arts college trying to understand adjudication and its relationship to water users in New Mexico. A simple summer research project, I say. Hector snorts.

At that moment, the tension between us feels thicker than wet clay. For Hector, adjudication is a four-letter word, the *A* word. The state of New Mexico uses water adjudications to map and redefine water access as a private-use right, not bound to the village or valleys, parsed to individuals and not the communal institution of the acequia. My seemingly innocent question is perceived by Hector as an alien, unnatural framing of water in legalese. For hundreds of years, the acequias were largely left alone to allocate and manage water along their ditches. Hector's acequia is one of the hundreds of these local water sovereigns, as I have come to think of them, that preexist the state of New Mexico.

Water sovereignty explains much of Hector's reaction in this context. Redefining water in any cultural, political, and historical setting is contentious. Members of the ditch control water, which they think of almost as family. Thus, the intimate sovereignty over water is not perceived as only political. Sovereignty includes the lived practice of managing the water as an essential part of the cultural and sacred landscape, across hundreds of valleys in the state. Hector is concerned about losing control over decision-making, about local water governance, *and* about keeping water attached to the land.[2] Losing the water would imperil the sense of community, as he told me:

> Listen, Eric. No, we haven't been touched by that process [adjudication] yet, and a lot of people are nervous about it up here. Once it starts, it touches everything. Everything ... old family grudges, all the cultural stuff between Pueblos [Indians] and Hispanos gets dragged to the surface again ... It exposes everything and everyone. In some ways I wish it had happened fifty or sixty years ago, you know? Back when there was more water use and agriculture on these ditches ... now [shaking his head] ... I don't know what kind of water will stay in this [Embudo] Valley once the state engineer is done with us. There's nothing simple about it ... and [tapping on my chest with a finger], *you're gonna be sucked into this for more than a summer if you ask the right questions ... just like we will be.*

Over a century ago, the 1907 New Mexico water code created the Office of the State Engineer and charged it with a monumental task. The state engineer was to perform "general stream adjudications," accounting for all existing uses of water in every watershed in the state. That task continues today. The state maps out diversion points, land parcels with water rights, the first date of beneficial use of water, and the crops that are grown and their water use in acre-feet per year. The details needed and captured are painstaking. For those undergoing adjudication, the stakes are high: irrigators, agencies, and cities are under pressure to make full and visible use of their water rights. Water users scramble to find old historical documents related to their first-use dates or deeds of property ownership. Adjudication sparks a scramble for time, priority dates, and evidentiary proof to get one's full measure of water. Studying this process is also a monumental task. Hector was right. This research lasted much longer than that single summer, and I soon found myself, like Alice in Wonderland falling into the rabbit hole, swept into the dizzying maze of adjudication.

Part of me wishes the A word had remained distant and foreign to me because of its complexity and its reach into all aspects of water. Nevertheless, I found it too fascinating and revealing to ignore. These water adjudication lawsuits expose everything that is strange and contentious about western water law and water use: disagreements over use, local and expert knowledge contests, competing legal notions of water, the allocation of water rights by crop, arguments about water's purpose, and interstate disputes over water.

Under the 1907 water code, water abruptly became a state-owned yet privately allocated resource. Adjudication was the process by which the state would translate water access to a private-use right. In some cases, adjudications went smoothly and quickly. These were typically in sparsely populated areas with little water to allocate. More typically, however, the legal process has been adversarial, costly, and lengthy. Multiple generations of families have been embroiled in the same adjudication lawsuit, and the most difficult and massive water cases have not even started yet. Water cultures in New Mexico, like the Hispano irrigators or Native sovereign nations who think in more collective, not individual, water terms, contested the state's rereading of their water norms and customary understandings of the purpose of water.[3]

Allocating small and big water shares will be an increasing challenge in this drier, warmer, more contentious century. Hector knows this, and so does the state engineer, but they think about water in different ways and at different scales. Scholars have recognized the importance of adjudications. Earlier contributions in a 1990 special issue of the *Journal of the Southwest* highlighted the problematic social tensions of adjudication as a process.[4] Those concerns remain thirty years later. Ten years ago, in an interview by Jack Loeffler, Frances Levine wrote, "No contemporary issue is as emblematic of the struggle between traditional and modern lifeways as the water rights adjudications currently under way in much of New Mexico."[5]

In the field of critical legal studies, scholars have demonstrated that multiple customary and formal legal traditions can coexist in the same space. New Mexico represents such a case where competing worldviews and customary traditions of water use endure to this day.[6] Legal scholars have assessed and critiqued various state approaches to water adjudication, often focusing on the expense, the lengthiness, and the legal dilemmas. These works were aimed primarily at audiences of water law professionals.[7] In addition, work by social scientists has addressed the impact of changes to water governance, water privatization, and the urbanization and commoditization of water.[8] Water infrastructure, changes in water law, and federal Indian water policies in the American West have also been well documented.[9] Historians too have provided rich accounts of the transformation of the American West's rivers, the impacts of large-scale irrigation and dams, and the movement of water to cities.

Unsettled Waters fills an unexplored space in the water literature, focusing on lived experiences of New Mexicans. I critically examine how adjudications affect water users, how they create new forms of water expertise, and also how they might be useful in addressing twenty-first-century water challenges. Adjudication spans generations, is ongoing, and has no end in sight. As I hope will be clear, adjudication has consequences, intended and unintended, for all water users. Here I focus on how the state translates and transforms water from a shared, necessary communal good into a singular resource to be owned by individuals.[10] From a theoretical stance, tracking adjudication allows us to examine how a state "sees" water and attempts to redefine its new water citizens as property holders. Adjudication transforms water into a private-use right through law, a system under which water becomes a potential commodity.[11]

COLLECTING NEW NARRATIVES OF ADJUDICATION

Unsettled Waters is based on a mixed-methods approach combining archival, field, and ethnographic research. Between 2006 and 2017, I conducted 274 interviews. Of these, 211 were of rural irrigators, with a special emphasis on those who belong to acequias or are in irrigation districts where acequias are present.[12] Local expertise preexisted the rise of disciplinary water "experts" (attorneys, engineers, etc.). I did not focus on a single basin or valley. Rather, I interviewed water users from basins around the state to get a fuller picture of this statewide process (see map 1).

I also interviewed sixty-three lawyers, engineers, historians, technicians, and water managers who were working for state agencies and in private industry. I included these informants because no single body of water users can claim a monopoly on understanding water problems, much less solving them. These water "professionals" also provide balance against an overly localized and nostalgic view of water in the Southwest.[13]

Because this is a book that depends on the views and perspectives of *living* New Mexicans, informant names are pseudonyms. A few interviewees wanted to be recognized by their real names, and I honored their requests. In some instances, I modified characteristics of the person depicted or quoted to disguise the source

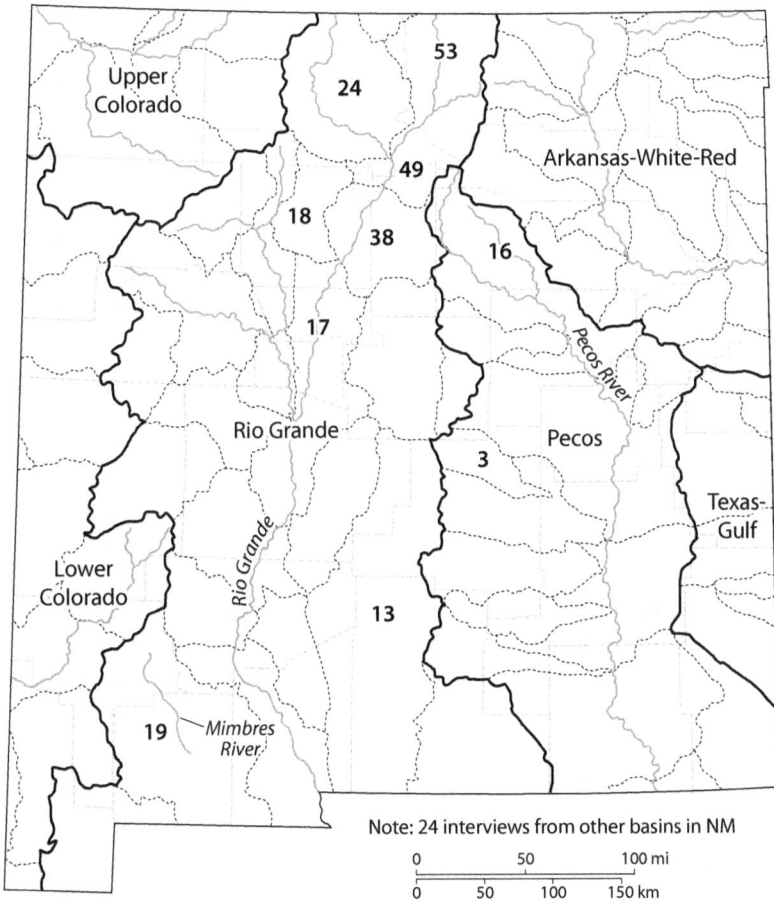

MAP 1. Interview locations included in *Unsettled Waters*. These are aggregated totals of the basin-specific interviews conducted by the author in New Mexico. Note that an *additional* twenty-four interviews came from smaller basins that are difficult to represent independently due to the map scale.

of information. I did not want to betray the confidences of those who shared sensitive, personal, or ditch-wide perspectives. When real names were published in the public record and legal documents, I used those real names. None of those cited, mentioned, or acknowledged bear any responsibility for misinterpretations of fact, fiction, or their own words. The views and voices herein reflect the concerns, thoughts, and constructive critiques of adjudications by New Mexicans.

Each of the interviews, stories, or accounts in this book has deep historical roots. Legal and historical archives were consulted and used to enrich my accounts of past adjudications, since many of these court cases have lasted through two or more generations of New Mexicans. Archival records were vital supplements to the gaps of peoples' memories as they recounted their court experiences.[14] Adjudication relies on court decrees and data, hydrographic surveys, maps, and charts. I consulted these resources in interpreting the regional cases that appear later in this book. The maps themselves, often dating back to the days of hand-drawn ink on linen, are gorgeous objects left behind by the technicians doing the field mapping. However, numbers and maps alone cannot provide a full picture of the process or its effects on those involved. Since the state, hydrologists, and other experts already track quantitative aspects of adjudication and state knowledge, this book focuses on the qualitative and cultural impacts of an unfinished process.

Listening to those affected by the process reveals how water users have questioned and contested the state's simplified reading of water as property. Following Freyfogle's treatment, private property has always been a contingent relationship, not a solid and identifiable "object" of property.[15] Water is owned by each state as a public good, but the use rights are private and dispensed to individuals by the state. Water rights are especially contingent since they depend on state-driven framings of water as a public good *and* an actual supply of water to use. Water itself in the West is thus a hybrid good: publically owned yet privately dispensed for use as a property right to use. This critical ethnography of water adjudication holds implications for those affected by the process, the state agencies and individuals doing the work, and those who have yet to be visited and adjudicated by the state.[16]

GOALS AND ORGANIZATION OF UNSETTLED WATERS

This is a hybrid text that uses the pragmatic lessons from New Mexican water users and experts and draws on insights from scholarship on water issues. I have intentionally written this as a kind of public political ecology, with as little jargon as possible.[17] There is much to learn here from the region and its people, which can inform the scholarly water literature and water policy in the West alike. The lengthiness of the adjudication process may actually have some benefits. There is still time to reform or adapt adjudication, and lessons from New Mexico extend beyond the state line. Few western states have completed their water adjudication processes, and all are seeking solutions to water scarcity and allocation challenges.[18]

I have divided *Unsettled Waters* into thematic sections. Part 1 focuses on the work of adjudication and case studies. Chapter 1 describes the roots and purposes of adjudication and how adjudication is linked to prior appropriation, as well as how both complicate cultural understandings of water in New Mexico. The two regional cases in chapters 2 and 3 exemplify how adjudications founder in basins with multiple cultures of water. Chapter 4 then details the problematic social, political, and hydrological consequences when adjudications leave the courts to become negotiated water settlements.

Part 2 examines what adjudications and settlements produce. Chapter 5 examines how adjudication has produced new metrics of space, time, and volumes of water. The adjudication-industrial complex has also produced new forms of expertise, as I argue in chapter 6. In chapter 7, I describe how new water-user organizations and regional water-planning strategies have emerged as by-products of adjudication.

Part 3 focuses on the future of adjudication and coping with new water demands and potential lessons. Chapter 8 discusses what threatens to be the hardest work of all: adjudicating heavily populated regions along the Rio Grande. Chapter 9 addresses climate change, the water demands of other species, and how to account for water in our new era. Finally, in chapter 10, I revisit the experiences of New Mexicans and how they may inform other western states struggling to count and allocate their waters.[19]

PART ONE

Unsettled Waters

*How Water Adjudication Works, What It Does, and
What Happens When It Fails*

1

How Local Waters Become State Water

Miguel never understood the logic in water adjudication. In his late fifties and a retired employee of the Los Alamos National Laboratories, he was now a constant gardener. Most of his concerns were for the younger generation along his ditch and those few people under the age of thirty still living nearby. Sitting on a lawn chair in the shade of his backyard apple tree, he reflected on adjudication's implications for him and his neighbors.

> I mean, I get that we have to know how much water we have, right? That makes sense, so that Texas doesn't get it all (he smiles a bit). But beyond that, what do we get out of this whole thing? They haven't even done my valley, and now they're warning us that the adjudication is coming to us soon, and we're not ready. We haven't organized yet like the Taos folks. My neighbors don't seem to be worried or alarmed, but they will be once it's here. Once the state engineers show up, it's all over, and it'll be too late for them to make any claims about having irrigated this or that patch, and then that water number gets fixed, and it's done. There won't be any future ability to expand water needs, I think. That's what no one here tends to get— once the process is over, *you don't get another chance,* and the amount of water we are using at the time of the process means that is the water we get, assuming no one sells their water or goes out of business . . . Then the engineer can figure out if there is any water we aren't using and then have that available for sale if there's some left over. It [adjudication] will change everything, even if people want to pretend that it won't change anything.[1]

ACEQUIAS AND THE HISTORICAL ROOTS
OF LOCAL SOVEREIGNTY

Like most of his neighbors in Rio Lucio, a small hamlet outside Picuris Pueblo, Miguel's home sits along an acequia, which provides the water for his small agricultural plot. It is a shared ditch with the nearby Indian Pueblo as it crosses through both indigenous fields and lands occupied by Hispanos like Miguel.[2] Acequias are gravity-fed irrigation ditches and institutions that were brought from the Iberian Peninsula when Spain was ruled by Moors. Acequia as a word has Arabic origins, meaning *water carrier* in its original form.[3] These institutions moved with the Spanish to Mexico and eventually to New Mexico during the Spanish Colonial period (1598–1821).

Notably, New Mexico underwent two major episodes of settler colonialism and three political shifts, starting with the Spanish and shifting to brief Mexican rule (1821–1846) and finally ending with US governance beginning in 1848. This resulted in complex political, legal, and cultural overlays and understandings of natural resources, including water and its governance.[4] Well into the twentieth century, acequias were the scale of daily water use, water governance, and life in New Mexico. Hundreds of these ditches exist across the state and are especially common in northern New Mexico, where there is greater water availability (see map 2).

Acequias as institutions function with the aid of a water boss (mayordomo), three commissioners, and the individual members of the ditch who use and maintain the ditch (known as *parciantes*). These institutions were vital to agrarian life and livelihoods in New Mexico's semiarid valleys. Today, the dependency on agriculture has decreased, but hundreds of acequias still remain functional. They are microdemocracies unto themselves, functional governing units of the state. They survive because they work.

Acequia members like Miguel and Hector (whom we met in the introduction) understand that water is work. It is work to be shared, via direct labor and through annual financial dues, before the water can be allocated. Members contribute to the annual spring cleaning of the ditch, known as *la saca* or *la limpia,* and to its upkeep. On most ditches, mayordomos coordinate with commissioners and other ditch officials on the same stream to estimate when all members on the stream system can begin irrigating and how much water might be available that year based on snowpack. Estimates are adjusted weekly and sometimes daily as fresh rain and snow events occur. This is a highly adaptable and responsive system that works with the actual amount of flowing surface water available rather than stored waters behind a massive dam.

In good years, parciantes can access the water when they need it for crops, gardens, and livestock. When drought or scarcity strikes, the hard part of a mayordomo's job begins: allocating water by shorter time rotations and watching

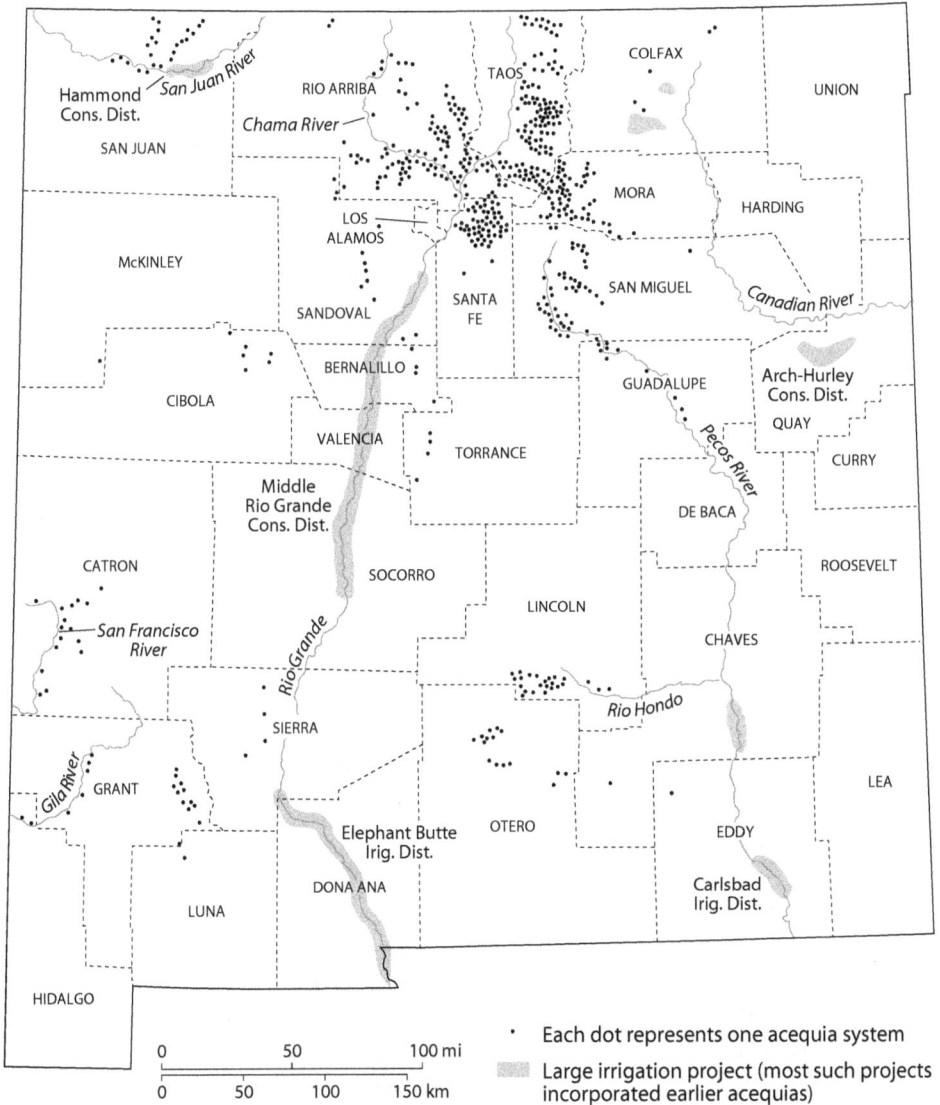

Hammond
Cons. Dist.
San Juan River
SAN JUAN
McKINLEY
RIO ARRIBA
Chama River
LOS
ALAMOS
SANDOVAL
CIBOLA
BERNALILLO
VALENCIA
Middle
Rio Grande
Cons. Dist.
CATRON
San Francisco
River
SOCORRO
SIERRA
Gila River
GRANT
Rio Grande
Elephant Butte
Irig. Dist.
DONA ANA
LUNA
HIDALGO
TAOS
SANTA
FE
TORRANCE
COLFAX
UNION
MORA
HARDING
SAN MIGUEL
Canadian River
GUADALUPE
Arch-Hurley
Cons. Dist.
QUAY
CURRY
DE BACA
ROOSEVELT
LINCOLN
CHAVES
Rio Hondo
Pecos River
LEA
OTERO
EDDY
Carlsbad
Irig. Dist.

0   50   100 mi
0   50   100   150 km

· Each dot represents one acequia system

Large irrigation project (most such projects
incorporated earlier acequias)

MAP 2. Distribution of acequias in New Mexico. Adapted from Utton Center (2013).

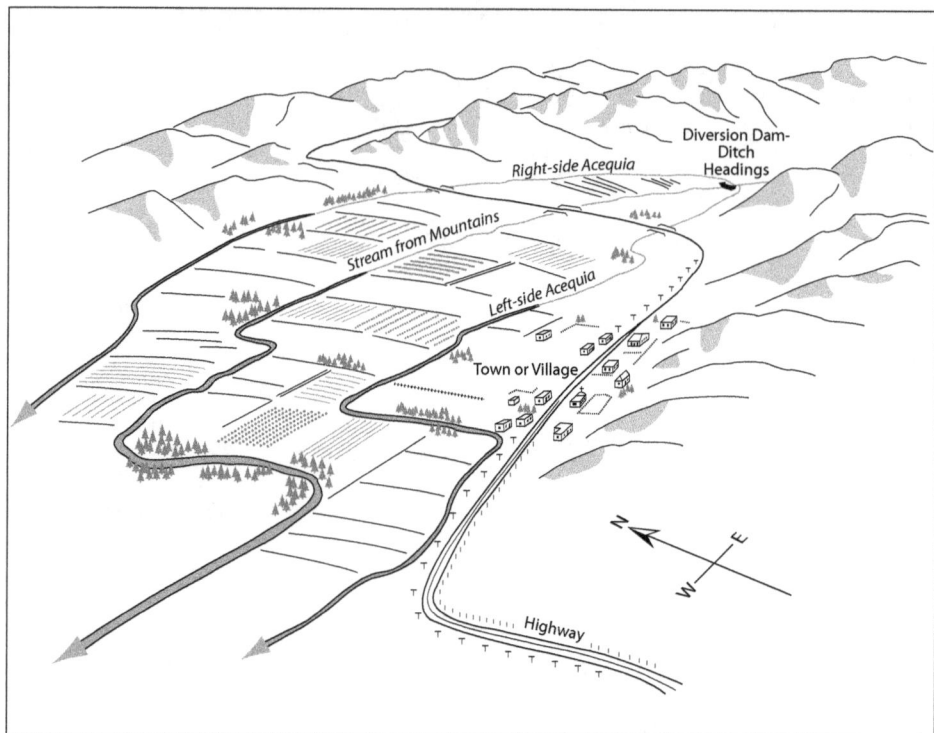

FIGURE 1. A hypothetical valley in New Mexico with acequias. Adapted from Utton Center (2013).

individual water use to ensure all members' needs are met. The mayordomo designates when and how much parciantes can irrigate and monitors the water flow. Access to the ditch can be blocked if parciantes fail to pay their dues or take water out of turn. This is an important point: The institution of the acequia—run by the commissioners and mayordomos—controls access to and use of the ditches that carry the water to which individuals have rights. In other words, acequia rights are not the same as individual water rights. Parciantes have to follow the rules of the acequia institution to keep their access to ditch water and maintain their individual water rights.

Acequias as physical features extend the riparian habitat, stitching together patches of emerald floodplain that weave through dry hills dotted with piñons, junipers, and cacti. Ditches can be on one or both sides of a diverted stream (see figure 1). They serve to both widen the floodplain and to store more water underneath the upper watersheds for longer time periods. Acequia landscapes are deeply altered. The ditches were built and are maintained for human use. They benefit agriculture at the expense of natural stream flows and have consequences for fish as well as mammals such as beavers and muskrats.

Long before New Mexico existed as a state, acequias were the essential institutions and objects of Hispano community formation. Pueblo Indians adopted the ditch and institution system. The overlapping cultural and historical layers surrounding these ditches and their impacts on landscapes and communities make New Mexico fascinating and often distinctive compared to other western states. The state of New Mexico's 1907 water code, written when the region was still a US territory, imposed a new set of water laws and water-user expectations. The reverberations of new and conflicting water regimes still resonate in Miguel's village.

Miguel expressed suspicion, even fear, about upcoming water adjudication. Many other interviewees, especially those of indigenous and Hispano descent, echoed his sentiments. Such feelings are associated with the historical use of this word *adjudication*. Whether accurate in the long term or not, many New Mexicans fear that adjudication signals the final dismantling of the Hispano moral and communal economy of water. This perception has logical roots considering the historical record of land and resource dispossession in New Mexico.

Miguel put it bluntly. "Look," he told me, anxiety clear in his tone and face. "First the US came for us as people, then they took our land grants, then they took away access to our forests, they took our animal grazing permits, and now a lot of us think they want our water." It was the *A* word, adjudication, that stripped Spanish and Mexican-era land grants away from Hispano villages more than a century ago.

Just after the Treaty of Guadalupe Hidalgo was signed in 1848 ceding Mexican territory to the United States, plans and bills to incorporate the Southwest into the union were well underway. During the US territorial period (1850–1911), vast tracts of the old Spanish and Mexican community land grants were dissected through legal procedure, as well as through graft and outright fraud. By the 1920s, over 90 percent of community land grants were dispossessed.[5] The loss of land has not been forgotten in New Mexico and resurfaces in debates regarding water adjudication. Other losses to resource access compound the distrust, such as the limits on forest use and plant collection and the loss of grazing permits in forested regions. These losses and restrictions particularly affected female property owners.[6]

One of the early US congressional bills to receive consideration in 1849, just after Nuevo México became a territory of the United States, proposed that a commission be established to adjudicate all lands and gold claims in both New Mexico and California. Even Senator Benton from Missouri, a proponent of manifest destiny, had deep reservations about forcing New Mexicans to beg and plead in a commission setting for their rights. As he argued to the US Senate:

Two hundred and fifty years have elapsed since that country was granted to its conqueror, Don Juan de Oñate: almost ten generations have lived and died there. Yet they are all to be called upon now to show their land titles, and to prove them also, back to the time of the conquest. All titles are to be ripped up, and rooted up, back to the original grant, two hundred and fifty years ago. What would Virginia say if she had been conquered by a foreign Power, and should be served in the same manner?[7]

Benton was no champion of Mexicans (or New Mexicans), yet even he saw the lurking danger of this federal move to humble new citizens before a kind of quasi-administrative and legal court proceeding.[8] Few Spanish and Mexican land grants were ever confirmed or honored. However, that does not mean their traces have disappeared. In rural communities, families with long roots vividly remember the fraud and shady legal procedures by which their lands were taken. Given this history of dispossession of communal land and resource access, suspicion about water adjudication is understandable. It is not just the state asserting its territoriality over water that is of concern. More central is the state's insertion into local and regional water governance.[9] New Mexicans worry about the delocalization of water and the potential loss of collective water sovereignty.

For Miguel and his neighbors, the concern is less with the awarding of a water right, per se, but rather with what individual rights might allow: a potential and future sale of those water rights away from their shared ditch. In the new system of water law established in 1907, a privately sold water right would no longer be used on the adjacent land. It would "leave" the ditch's institutional control, leaving less water for the community to work with in the long term.

Adjudication also has institutional implications on the ditch. Adjudication certifies an individual's water rights, not the community ditch water rights under Spanish and Mexican norms of water distribution. Acequia members fear that adjudication will endanger their closely controlled water commons, where participatory labor, citizenship, and water use are tightly conjoined. The results of adjudication create a patchwork of private water rights owners who are then less tied to the community that built the ditches. These private owners could also theoretically dispose of their individual water rights as they see fit—to their economic benefit but potentially harming the communal aspects of the ditch as a whole.

Adjudication quietly sorts a water-use right as a private-use right, abstractly moving water out of the realm of communal and acequia institutional control. In doing so, adjudication reveals much. I have come to think of this in geologic terms, which, given the pace of adjudication, seems apt. Geologists study stratigraphic layers to understand deep time and Earth's formation. In adjudication, the state is seeking to extract a single "core" meaning to water, a fixed amount to be bestowed as a private-use right. To get to that layer, the state bores down through time, through the different cultural values given to water and the historical disputes between cultures. Looking over the state's shoulder, peering down this conceptual borehole, studying adjudication and its results, illuminates those past layers of cultural water: the communal water in acequias, the shared indigenous waters of the Pueblo, and the individual water rights of later settlers and city utilities. To enrich political theorist James Scott's thoughts on the "state visioning" of water users, adjudication has multidimensional side effects: it *reveals* old water conflicts and *produces* new adversarial relationships between water users.[10] Adjudication litigation as a singular state coring of water activates the multiple (vertical) layered definitions of water, even as the state continues to insist on a singular horizontal, two-dimensional private-use right for water users.[11]

For community ditches across the state and for indigenous sovereign nations, there is fear that water is being translated as only a private good, as *only a resource,* so that others can locate and buy it as a "commodity." These competing sovereign views of water, and cultural identity, not only reflect the multiple waves of colonialism in New Mexico. They also reflect how modern capitalism has rolled out in this state, severing ties between communities and the surrounding landscape, resource by resource, layer by layer. For so many people that live along traditional ditches, putting individual price tags on natural features of the landscape can seem like madness.

Forests, land, livestock, and now water are all subject to new commodity definitions. Like the peeling of an onion, the old land grants created initially to serve Hispano villages were separated into redefined layers of timber, real estate property, animals as commodities, and "water resources," or "water rights," that can be made profitable and mobile throughout the state.[12] Fundamental cultural, political, and economic differences clash in the process: the perception of water as a communal shared good versus the new political economy of a private-use water rights system imposed by the state of New Mexico.

Adjudication seems perfectly harmless to the state and its employees. After all, they just want to seek out and certify New Mexicans' water uses as private-use rights. To the prior water sovereigns, on the other hand, the process can seem ominous as the multiple understandings of water are "cored" by the state in a single and simplifying way. For people like Hector or Miguel, this state redefinition of communal water into individual water rights is a violent one, even if that violence is slow, gradual, and often invisible.

It is important to note that water rights can be sold prior to adjudication or even during an ongoing adjudication.[13] Adjudication is the state recognition of individual usufruct property rights, not an automatic pathway to selling water. This means that landed property owners who have water-use rights can choose to sever their water rights from the land if, for example, they choose to stop farming and irrigating their lands. Then the water, priced per acre, is no longer just a private-use property right but has become monetized, a commodity that can be transferred. The amount of money paid per acre has everything to do with location.

A nearby city interested in acquiring water rights might pay up to $50,000 per acre-foot. If the farm is in an isolated rural setting, the price will often be half or a third as much.

The legal process does make those water rights more visible to potential buyers. The state itself is not commoditizing water per se—it is simply mapping, accounting for, and creating an inventory for water rights across the state. The Office of the State Engineer (OSE) does individuate and locate that private water right in time and space, by crop duty for the amount of water per acre needed or used, allowing for future marketing of water. Attorneys, water bankers, willing buyers, and water-rights owners then mobilize that water market to price the water itself. From the state's perspective, the 1907 water code was simply created to affirm and map individual property rights as a neutral process. State officials I spoke with were often frustrated by local perceptions of adjudication. Nevertheless, state technicians and attorneys should understand that these suspicions and attitudes are based on repeated experiences of past resource access losses.

Antonio from Truchas expressed a common concern regarding potential water transfers. "Losing our water from this ditch would leave a deep cultural wound that we'd never recover from," he said. His fears may sound extreme, but they are not unjustified. Water moves across basins in the contemporary western United States. This suspicion about making waters nonlocal, held by multigenerational New Mexicans, is often unintentionally confirmed by engineers, attorneys, and state engineer officials. Attend any public meeting on water in New Mexico (and elsewhere in the United States) and you will hear water experts and housing developers calling for water to be put to its "highest economic use." What they mean is for water to be moved from X function to Y function so as to generate greater economic value per acre-foot. When farmers or rural residents hear this, what they hear is "let's get water away from farmers and ditches . . . and get it to the suburbs, the city, industries, or more suburbs." This neoliberal and triumphalist free market rhetoric confirms the worst fears of farmers as just another way to put a price on water and move it to cities and industries.[14]

Miguel, Hector, and most other irrigators understand that adjudicating water rights and potentially selling water rights are two different things, yet they see them as intricately and sequentially linked. Hector told me about a visit from a county tax assessor to stress his point. In Hector's recounting, the assessor claimed, "I'm not here to raise your taxes. I'm just valuing what you have. Taxes are set by the county commissioners, not me." Hector answered this assessor with, "Yeah, but if you don't raise the assessment, then my taxes won't go up, right?" Thus, in a similar vein, Hector sees adjudication as the flywheel for pricing, selling, and ultimately moving water across basins.

Every potential sale away from the acequia, from the village, would mean less water for local use. Selling water rights away from the community does not just move the water; it erodes the basis for a shared water community that has under-girded many small villages across New Mexico for generations. The push to for-malize water rights at an individual level can have serious consequences for other kinds of property arrangements that are community based.[15]

THE PRODUCTION OF STATE WATER

Mandated in the 1907 water code, water rights adjudication in New Mexico was designed to map all perfected (in use) water rights. Water was declared state-owned and public, yet the individual-use rights to water would be privately held. New Mexico does not distinguish or rank the order of beneficial-use categories among agricultural, urban, and industrial and treats all water uses equally. *Beneficial use* of water is the basis for water rights, the measure for water rights (based on amount of use), and the limit (maximum award) for the awarding of water rights and is rooted in prior appropriation law. The presumption is that beneficial use has economic benefits, although this definition has become more expansive in the twentieth century and often includes recreation and instream flows. Prior appro-priation also established a historical ordering of first-in-time, first-in-right for the use of water: the earlier the use, the better (or more *senior*) the water right.

In more than a century since 1907, only about a dozen basins or subbasins have reached the final decreed stage of adjudication and are considered "complete" (see map 3 for completed and pending adjudications as of 2017). Most early adjudica-tions, between 1910 and 1950, were executed in basins with low populations and with few Hispano acequia claims or unquantified Indian water rights involved. What remains to be finished is daunting.

The state water code charged the OSE with conducting so-called general stream adjudications. *General* is a misnomer. *Universal* might be a better descriptor. The work is specific and meticulous and, by design, not particularly efficient.[16] Given this massive task, it is understandable why adjudication took so long to begin and why it is still ongoing. Since its inception, the agency tasked with adjudication, the OSE, has struggled with low staffing, underfunding, and the scale of the process. Adding resources and personnel is difficult. New Mexico is one of the poorest states in the country, and as the former head of the legal division at OSE, D. L. Sanders, put it in 2006: "No Governor wants responsibility for making government larger."[17]

New Mexico's water adjudication process consists of seven general steps (shown in figure 2). These legal suits are prepared by the OSE and then triggered in concert with the state's attorney general.

Each stream adjudication is more complicated than the diagram shown would suggest, and many of the steps in each phase are revisited multiple times. While every adjudication is unique, all include the three main phases shown in figure 2: the research and hydrographic work, the "subfile offers" to individuals, and the

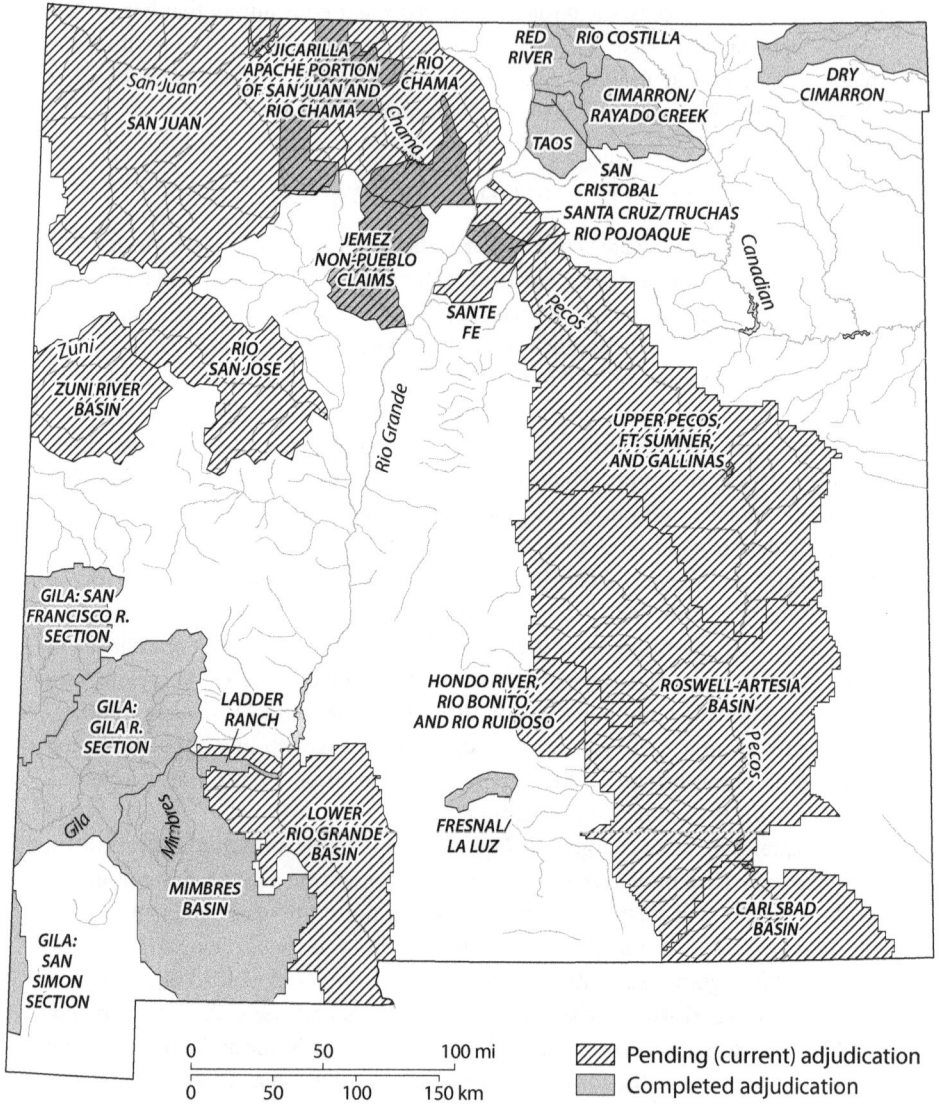

RED RIVER
RIO COSTILLA
DRY CIMARRON

JICARILLA APACHE PORTION OF SAN JUAN AND RIO CHAMA
RIO CHAMA
San Juan
SAN JUAN

CIMARRON/RAYADO CREEK

Grande

TAOS

SAN CRISTOBAL
SANTA CRUZ/TRUCHAS
RIO POJOAQUE

JEMEZ NON-PUEBLO CLAIMS

SANTE FE

Pecos

Canadian

Zuni
ZUNI RIVER BASIN

RIO SAN-JOSE

Rio Grande

UPPER PECOS, FT. SUMNER, AND GALLINAS

GILA: SAN FRANCISCO R. SECTION

GILA: GILA R. SECTION

LADDER RANCH

Gila

Mimbres

HONDO RIVER, RIO BONITO, AND RIO RUIDOSO

ROSWELL-ARTESIA BASIN

Pecos

LOWER RIO GRANDE BASIN

FRESNAL/LA LUZ

MIMBRES BASIN

GILA: SAN SIMON SECTION

CARLSBAD BASIN

0   50   100 mi
0   50   100   150 km

▨ Pending (current) adjudication
▨ Completed adjudication

MAP 3. Locations of completed and pending stream adjudications in New Mexico as of 2017. Dark shading with cross-hatching, like the Jemez River Basin, designates areas where all non-Indian claims are filed but Indian water rights have yet to be determined. Adapted from New Mexico Office of the State Engineer map sources.

Phase 1
- Attorney General files for basin adjudication in coordination with the Office of the State Engineer (OSE)
- Claimants (water rights holders) are sued, enjoined to the basin adjudication lawsuit
- OSE conducts hydrographic survey (technical phase)

Phase 2
- Subfile phase (letters of offer are sent to individuals)
- Streamwide issues phase (legal claims to streams that may affect all parties)

Phase 3
- Errors and ommissions phase (cartographic, legal, diversion or water duty changes)
- The *inter se* phase (individuals get to contest each others' water rights)
- Entry of the final adjudication decree in court

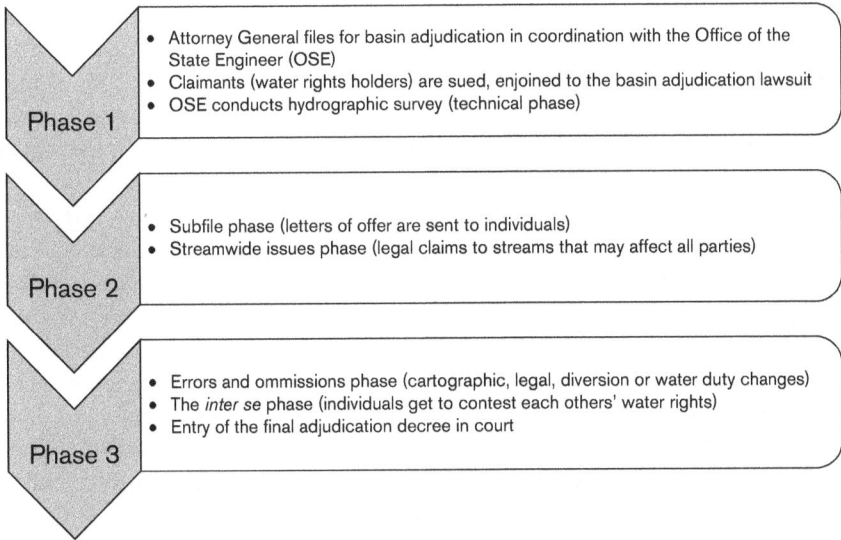

FIGURE 2. Simplified flowchart of the water rights adjudication process in New Mexico. Adapted from the New Mexico Office of the State Engineer and *WaterMatters!* (Utton Law School: University of New Mexico, 2013).

larger between-parties inter se process, whereby individuals get to question the rights claimed by other water rights users. While difficult, the technical and research stages of adjudication (in phase one) take far less time than the more contentious courtroom-based procedures on water rights between the parties.

WHAT TRIGGERS ADJUDICATION

Most general stream adjudications in New Mexico start in one of the following ways. First, adjudication can be done according to figure 2, with the state's attorney general and state engineer filing the case. Second, adjudication can also be triggered by any water rights holder whose claims have been recorded by the OSE in the basin. Third, large water infrastructure projects can also prompt the state to quantify water rights if those water rights would be affected by the new dam, canal, or pipeline. In contrast to other western states, such as Colorado, in New Mexico it is usually the state that files the complaint starting the general stream adjudication process, with the OSE and the attorney general working in concert to file the necessary documents. This is the first adversarial aspect of adjudication in New Mexico. Each step, or *phase,* of the adjudication process can take years to decades to complete. Naturally, delays ensue if claimants in the basin sued by the state do not respond to *offers of judgment* in phase two of the process. In most cases, phase three of the inter se process, in which water rights claimants can contest each other's water rights, is what takes decades. This is the second—and more problematic—adversarial aspect to adjudication, as it complicates social relationships between water users. But each phase can be fraught with complicated cultural and historical baggage.

Nearly seventy-three thousand defendants are now enmeshed in basin lawsuits to ascertain their water rights, in twelve pending adjudications. By one recent estimate, half of all water rights holders are now in view by the state agency (OSE), even if only 20 percent of the state's basins have been fully adjudicated.[18] The disadvantage to New Mexico's meticulous approach is that it takes so long for the state. The advantage is that the process is thorough enough that the state, when finished, will have a fairly accurate view and quantitative understanding of how much water is claimed, allocated, and used in any given year. This will, in theory, make it easier for the state engineer to conduct priority administration of waters and manage water allocation.

The length of time to complete adjudication is not always tied to the size of the basin, the amount of water, or its complexity in a biophysical sense. Notably, since the early 1980s, most adjudications that have included Indian water rights have ended not in state adjudication courts but in what is known as *settlement,* a less court-driven but no less expensive and complex process. These agreements and water settlements are meant to make the process less adversarial, but they often come at a huge cost, as I discuss in chapters 2 through 4. One cannot write about water rights adjudications without discussing water settlements, especially in a culturally diverse state like New Mexico, which has so many sovereign indigenous nations.[19] These agreements and settlements are ways in which local water users can renegotiate the state's power. They find new ways to not be governed by the state, rather than have the terms set from court litigation.[20]

TIME, WATER RIGHTS, AND ADJUDICATION

Water rights hinge on the *provable* date of first beneficial use. If all the rights under consideration in a basin are later than, say, the establishment of New Mexico as a state in 1912, the task is easier. The 1907 water code firmly established the use of prior appropriation law (first in time, first in right), common throughout the American West, which supplanted previous customary traditions. Influenced by Colorado and that state's strict adherence to prior appropriation, politicians at the time agreed to this template of water law for the state of New Mexico.

Under prior appropriation an individual with an earlier date, say, 1730, as a first-documented diversion and beneficial use of water can get a full allocation of water rights before those with more recent use dates. As long as those water rights are used continuously on appurtenant land or at least not discontinued for more than five years, those water rights "stay" with the adjoining land.

Pre-1907 water rights, which preexisted the New Mexico water code, are recognized as senior water rights. These include both Pueblo (Indian) and Hispano water rights, which get further distinguished. If the entity is a sovereign nation, such as a Pueblo Indian land grant or a Navajo reservation to adjudicate, then US federal government agencies like the Bureau of Indian Affairs or the Department of the Interior, in their trust relationship with Native American nations, step in as a party to the suit. It gets more complicated. Some Indian water rights (typically, Pueblo) are awarded on a historical irrigation acreage basis, using archival and archaeological support, whereas others (like the Navajo or Apache) can be awarded based on what amount of land might be irrigable, referred to as *practicably irrigated acreage.*

If Hispano post-1598 water rights are at stake, the adjudication also involves extensive archival research conducted by state engineer personnel or contract historians to prove first-use dates by various individuals along the ditch. Acequia members often insist on the communal notion of shared water-use traditions that are the norm on these ditches. They often argue for a single date for the entire ditch (acequia) instead of differential and individual water rights dates. Prior to adjudications, most senior water users ignored strict prior appropriation, especially on Indian lands and along acequias. In both cases, the basis for allocating water was based on the amount of land held, equitable water sharing, and demonstrated need. Water allocation, sharing, and disputes were already complicated enough before the 1907 water code was established.

All of this makes for a byzantine water world. It is perhaps no wonder that scholars outside of law schools have ignored these state procedures.[21] On the state's end, one of the very reasons that adjudication is "so damn slow," as one attorney put it, is because of defendants and their reluctance to engage with the state engineer. There is fear, anxiousness, and often resistance to responding to letters and paperwork. They do their best to ignore the OSE. As a long-time resident of the Embudo Valley put it: "People just like to ignore the state engineer . . . It's part of a long history that we just don't trust the bureaucrats and engineers making decisions about our water."[22] This is not just a refusal of state authority; it is a refusal to acknowledge the state itself.[23] Privately, some irrigators insisted the state has no right to fraction out water rights on their ditches, which is partly why they are reluctant to share information with the OSE. This tactic, however, is only effective at dragging out the time line of making offers to water rights claimants. If the OSE does not hear back on offers of judgment, adjudicating personnel move ahead, certify the right as complete, and assume that their own state historical research on that person's individual water right was correct. These dating exercises over time, ditch, and priority play out to full effect later on when the OSE has to conduct priority administration in times of drought.

Finally, federal water projects, such as dams, force the OSE to adjudicate water because of the affected water rights. To use a computer analogy, water law is the "software" that influences the structures (engineering) that need to be put into place. The "hardware" of engineering then shapes how water law works on a more practical level once placed on a landscape.[24] The water code of 1907 provided the software and legal basis for water priorities and allocation. To be clear, both federal and state dam projects require the start of adjudication to sort the stored waters by water rights holders.

Think of the vast quantity of water held behind a dam: that water is used by those who hold "rights" to particular amounts. Different people may hold different water rights to different sets of water within the same reservoir, whether "native" to that basin or transferred from another basin.[25] Fortunately for New Mexico—and most states in the American West—the large reserves of stored water created by dams delayed the need to strictly enforce prior appropriation. This is important to note: things could have been a lot worse without dams, in terms of water scarcity and senior "calls on the river" demanding the state engineer enforce prior appropriation.

Twentieth-century dams and reservoirs bought twenty-first-century water managers and state engineers some additional time and flexibility to adjust to new laws and water allocation technologies. The availability of reservoir water held behind engineered dams has largely allowed water users to avoid senior versus junior water conflicts and to not worry about the actual water law in the West—as tools for storing water, they seemed unmatched.[26] Enforcing prior appropriation law is not an easy or desirable process for the state engineer. In fact, prior administration of waters is jokingly referred to in the OSE offices as "the nuclear option." The slowing and vast storage of water changed both the timing of water releases and *mitigated* the need for prior administration of water by the state engineer. Massive infrastructure projects connected small farmers like Hector to state adjudicators, water managers, and urban residents across the state, as demonstrated in the next few chapters. Western states completely replumbed their rivers and streams. These new cultures of "expert water" and associated infrastructure were coproduced right alongside adjudication (see chapters 5 and 6).

LAWSUITS ILLUMINATE THE NEW MEXICAN LANDSCAPE

Adjudication as a process "was never meant to be personal," as one senior adjudicator put it to me in 2009.[27] Yet these lawsuits, filed by the state, call out defendants by name and thus feel deeply personal and adversarial. Some defendants' names become infamous. Imagine that your name is Enrique Abeyta, for example. You come from a long line of Abeytas, a rather common family name in New Mexico, yet now your name has come to suggest something different: a lawsuit. The first alphabetically listed defendant (Abeyta, in this example, or Aamodt for another) in a water adjudication suit becomes the name by which the suit is colloquially known. As one adjudicated farmer joked ruefully, "No wonder they haven't finished, all the adjudications start with the letter A, they never get to Z!"[28] Each of these family names is shorthand for the basins under adjudication across the state of New Mexico.

Irrigators, municipal water employees, and even local politicians were all eager to talk to me about watersheds and water management. However, utter one of the *A* names associated with a case and their reactions changed. The dead stare, the aloofness, the skepticism on peoples' faces were all clearly visible. Of course, it wasn't about the defendants bearing the unlucky names. It was what those shorthand nicknames now meant—the cases themselves and the uncertainty they brought. Even adjudicators get tired of the process over time. "If we didn't have to adjudicate, Eric, we wouldn't," a former adjudicator and OSE attorney told me, her voice and face betraying her weariness.

That said, adjudication was not always difficult, tiring, or perceived as problematic. In early test case basins with little cultural or legal diversity to contend with, the state engineer's personnel were not regarded with suspicion. After all, the technicians were simply there to map, document, and title out property rights that individuals had long claimed and used. Such was the case along parts of the Mimbres River and the Canadian River, which were done and decreed efficiently. These were less controversial because of the lack of legal pluralism and the lack of indigenous and long-standing Hispano claims to water. Fewer cultures of water translated to fewer hiccups in thinking about water rights as property-use rights.

Technicians who worked on the Mimbres adjudication recalled it as simple and straightforward. "It was pretty easy in the long run . . . maybe part of it was that there were no Indian water rights claims there, for sure, but overall [shrugs shoulders], it was a cakewalk compared to some of the other cases up in northern New Mexico where everyone seemed ready to question every damn thing we were doing in the courts, in the fields, and everywhere in between."[29] These sentiments were echoed in other interviews from the more culturally diverse basins that remained stuck for decades.

It was the foreignness of prior appropriation water law that struck many as odd. The state seeks to find dates of first use in time to match up to these last names and land parcels for an orderly hydraulic state. There is a logic to the state's approach in that prior appropriation water law is predicated on the "use it or lose it" basis of beneficial use. When state agents appear to map and quantify water use on a particular plot of land, it is in everyone's best interest to look or act as if they are fully using their claimed (or historically used) water rights. There's little incentive under prior appropriation to actually conserve water or to use it efficiently.[30] Water rights holders often think they have to fully use their water rights to keep them perfected. That is not the case. But because beneficial use is the "limit" to individual water rights, as explained previously, irrigators and managers alike have no incentive to go below the maximum limit of their beneficial use or their assigned water duties, depending on crops.

Adjudication was supposed to be a template process using watersheds as a basis for doling out water rights.[31] The new legal system and state code were nevertheless received as having to fully use one's water rights under prior appropriation ("use it or lose it"). Local irrigators and water users quickly understood what was at stake and carried over that understanding into the "performance" of water use when OSE personnel were completing adjudication maps for the home office. In basins where all possible cultural-legal understandings were all present, the performance of using water was even more critical. This came through in all my interviews. When adjudication was underway in the mapping phase, it was good to be seen irrigating, pumping, or diverting.[32] The 1907 water code did share much in common with Spanish Colonial practices for making good on property rights: the act of property possession was about visibly performing those rights and relationships between property owners and the authorizing agent.[33]

One water user from Taos, for example, claimed that "my neighbors started using a lot more water when the field mappers [from OSE] were here doing the maps and stuff."[34] Even in adjudicated areas that were less problematic, such as the Mimbres in southwestern New Mexico, an irrigator recalled that "that whole process [of adjudication] changed how we deal with each other; it added a lot of suspicion back then [1970s] that hasn't really disappeared. It's only made things worse in a dry area."[35] He went on to say, "Sure, the whole thing [adjudication] was wrapped in the late 1980s [1989], but we had no idea what it would trigger . . . in pockets of the valley, there were no issues, but in others . . . it just triggered bad blood and some civil suits, some of which just won't go away."

From the state of New Mexico's perspective, the cultural histories and geographies of water use do not legally matter until adjudications are underway. However, differing federal and state views of identity and bloodline can splinter water users, sometimes in the same family. These complexities of identity were simplified, often reduced to a binary of "Indian" and "non-Indian" identity and citizenship in adjudication issues. This unfortunate splitting of water identities is the result of federal policies and definitions of who is considered an Indian within the United States. It is a kind of biopolitics that does not mince on identity: either you are Indian or you are not, as recognized by US federal entities. Native sovereign nations now also control their own tribal registry rolls for membership. Assertions or assigned definitions of who claims to be indigenous, Hispano, or of mixed Indo-Hispano identity are fraught with challenges, and these binary cultural borderlands are patrolled regularly.[36]

Binaries of identity and membership can work for creating transparent govern-ance or rulemaking but can complicate cooperation in water matters in New Mex-ico. Federal versus state legal treatment matters greatly, dictating what kind of water right a person is entitled to according to his or her identity. Next, I present an example of this complex and often perverse cleaving, along with a later tale of two brothers divided by this water-identity issue.

IDENTITY, FEDERALISM, AND WATER SOVEREIGNS

Identity and history matter in New Mexico's daily water governance. In 1598, new Spanish settlers arrived near Ohkay Owingeh Pueblo (which the Spanish quickly renamed San Juan Pueblo), home to one of the many Pueblo Indian groups along the Rio Grande. The Spanish eventually chose the western bank of the Rio Grande near what is today Chamita and the junction of the Chama River and the Rio Grande, their first attempt at a new capital, to be called San Juan de los Caballe-ros. The next year, however, the new capital was moved east to San Gabriel. An early ditch was dug in 1598, but the colony of San Gabriel also did not last and was officially abandoned by 1601, although some Spanish and Tlaxcalan stragglers may have stayed behind. By 1610, most officials had moved to the new capital, Santa Fe.[37]

Those early ditches near San Gabriel were likely used by the Pueblo and any remaining Spaniards until the 1680 Pueblo Revolt. This is one of the reasons why it is difficult to claim and get awarded any pre-1680 water right in New Mexico: most Spanish documents burned in the revolt. Spanish colonists and indigenous peoples (from Mexico) returned to this area in the late 1690s, and their descend-ants have been there since. Consequently, the Pueblo and surrounding towns share ditches, bloodlines, and a complicated history of kinship and identity.[38] Most pre-1680 Spanish records were incinerated and lost to historical memory during the 1680 Pueblo Revolt. Because of this first wave of colonialism, the resulting cultural politics of water use were already complicated prior to 1846. Water politics based on identity were commonly defined by the *blood quantum* (percentage of Native descent) understandings of tribal membership, as con-structed by the United States federal government and now often enforced by Native sovereigns themselves.[39]

To add to the complexity, the Pueblo were incorporated into the United States as *Mexican citizens* under the Treaty of Guadalupe Hidalgo, not as indigenous peo-ples. Since they were settled at the time of US takeover in the region, they were not considered "wild Indians" who posed a threat. They also seemed to live like their Hispano neighbors, with established Catholic churches on their lands. The Pueblo Indians were not given federally recognized "Indian" status until the 1930s, and they were deprived of voting rights until that status was formalized (1948). This treatment of Pueblos-as-Mexicans-then-Indians continues to haunt New Mexican water rights adjudications and settlements. Indeed, New Mexico is one of the few states where the concept of the *colonial present* still means something, as the politics of identity "recognition" continue to operate and confound sovereign politics.[40] The problematic reversal of these identity assignments under federal Indian policy continue to have real policy implications (as I discuss in chapter 2). These matters have only become more complex over time under the second wave of colonialism that swept through the region after 1846.

The question of federal Indian identity and recognition is alive for residents of Chamita and the Ohkay Owingeh Pueblo. Because of shared canals running between the Pueblo and Chamita, cooperation has been necessary for hundreds of years. Irrigators from Chamita and the nearby Pueblo have sometimes struggled but found ways to come to agreement. To illustrate the complexity of identity politics and water sovereignty in New Mexico, I turn to a story of two brothers.

Juan Pacheco is a member of one of the Chamita acequias and has been active in its oversight and governance for decades. He considers himself Nuevomexicano, of joint Indian and Hispano descent, and actively participates in local affairs. His brother, Miguel, identifies as a member of Ohkay Owingeh Pueblo and has also been prominent in local and regional Pueblo affairs. These two brothers, from the same family, with different allegiances, are partitioned into different categorizations for water adjudications. Juan, defined as a non-Indian (as litigation calls all nonindigenous peoples), falls under state jurisdiction of the OSE and the rather rigid terms of prior appropriation. Miguel, as a member of Ohkay Owingeh Pueblo, has a different state-recognized identity as Indian and thus falls under different laws and jurisdictions. Federal identity categorizations (Indian and non-Indian), with added Native sovereignty definitions of who is formally enrolled as tribal members, define the brothers and separate their legal treatment and status into federal and state courts. Identity determines what kind of judge and court, state or federal, has purview in their water cases. Juan employs his Hispano identity in his appeal to the local and shared community norms of water management. As he put it:

> It's pretty absurd that we fight over this stuff and that we get really different legal treatments simply because of the cultural line or ditch we have chosen. I mean I'm as much Pueblo as my brother [Miguel], but because I didn't actively enroll as a tribal member, I get no federal backup, no federal representation, unlike my brother. It's ridiculous. And it means the state is effectively my boss; the state thinks it can just tell us what to do since we're under the New Mexico water code. My brother, he just laughs at this and says, "We don't recognize the state [of New Mexico] or its laws on water. We do what we want, and the feds will protect us. He's never that mean to me about it, but it's always there. He's just dismissive of the state engineer because they have the feds on their side.

Miguel, as a recognized member of the Indian Pueblo, falls under Pueblo (and thus federal) purview for water issues. Thus, Miguel leans on the definitions of Pueblo water rights.

> Since we've always been here as Pueblo, and first peoples, we fall under the federal jurisdiction and protections. It's not always great. We have had a lot of problems with the BIA and the [Department of] Interior officials, some bad attorneys along the way who didn't know what they were doing, but mostly it's okay. We don't have to worry about the state engineer [of New Mexico] because he has almost no authority over us. We usually just go about our business. It is difficult for my relatives on the other [Hispano] side since they fall under the state, and they have to listen to what the state engineer says and thinks, he tells them what to do . . . anyway, it is hard to work together because of this. We get along, but it could be better. We just don't want to be told what to do, or be forced into sharing water that we think is ours, that has long been ours, and that the acequia folks think is all theirs. You can't force a relationship or a compromise, right? You have to both agree to the terms, and we have our own cultural way of dealing and talking about water.

Juan and Miguel are thus well aware of the state-federal water divide cleaved through identity. Many Pueblos, and certainly individual Pueblo members, do not recognize the state's laws in managing, much less designating, water uses on Indian lands. Like the blood relationship in this example, water is shared across the Indo-Hispano communities. In interviews, I heard both Pueblo and Hispano water advocates using the phrase "since time immemorial" to highlight their respective rights and water use prior to the state's existence. These claims are a way to create a space of exceptionalism in either federal or state law, and both are a kind of state refusal.

Juan and Miguel exist in the same space but are bound to different water governance jurisdictions. They share the same DNA, but where they reside, physically and in political space, matters more than the actual percentage of Pueblo or Spanish lineage. The parsing of identity and resource rights alignment is germane to eventual administrative matters under state law (non-Indian) and federal law (Indian). Bloodlines shape the jurisdictional water rights of many in New Mexico, translating to how much water sovereignty each group can exercise. Acequias remain limited sovereigns, in charge of their local ditch water. The Pueblo and other indigenous groups of New Mexico have a federally protected limited sovereignty that is more regional in scope and broader than that of the acequias. Then there is the state of New Mexico with its presumption of state-public ownership of the waters across the state. Finally, there is the federal government with its trust responsibility for protecting both Indian water rights and endangered species.

The multiple concentric rings of water sovereignty complicate any singular notion about nation-state sovereignty and control over water. Sovereignty over water and the often-associated concept of water security may be passé when it comes to debating international water governance solutions. But here, rescaled to reflect local water sovereigns like acequias and the nation-within-a-nation indigenous sovereigns like the Pueblo and the Navajo, the jurisdictional aspects of water sovereignty explain why the state of New Mexico has struggled to redefine water and individual water rights.[41] These are all constrained forms of sovereignty, yet all of these water cultures exert some degree of water control and water sovereignty.

Sovereignty underscores the importance of identity in relation to water governance and local management issues. The 1908 Winters decision, for example, determined that Indian reservations have implicit water rights attached for future development. That court decision never offered a metric or quantification for reservations; there were no explicit guidelines. As a result, the implicit 1908 Winters Doctrine water rights were long ignored and left as a matter for later courts to specify and quantify. The elaborate construction of Indian identity by the federal and state governments—and how the category of "Indianness" has changed over time—has fed directly into problematic cultural water relationships. These place-bound identity issues cascade into fights over priority water rights by seniority and how water settlements treat this binary of Indian and non-Indian quite differently.

The Pueblo, along with other indigenous sovereign nations like the Navajo and Apache, barely acknowledge the state's power since they have nation-state federal protections in place for defending native waters.[42] OSE oversight begins and ends at their sovereign nation boundaries. The Native sovereign nations are skeptical about water rights being awarded by a state that exists largely at their territorial expense. Although the 1952 McCarran Act allows states to enjoin Native sovereigns and the federal government in state water adjudications, federal and state courts continue to stake out claims to parse out Indian water rights claims. Overlapping water sovereignty, culturally complex views, and claims to water all make adjudication more difficult.

In the end, who you are defined to be determines who protects or authenticates your water rights. In undeniable ways, the administrative view of individual water users creates differential citizenship for water resource governance. Adjudication has consequences that go beyond affirming liberal property rights regimes in western states. Adjudication is not simply about the transfer of ownership of property, or merely about water rights handed over to individuals. It is as much about managing water users in particular identity categories as orderly, disciplined state citizens, as Miguel's lead quote to this chapter suggests.[43] In the basins where Indian water rights are present, adjudication cleaves identity and creates a collection of both Indian and non-Indian water users (as treated by federal and state agencies and courts). Seeing like "a" state, then, is never singular in a federalist republic like the United States. Scott's approach was apt for critiquing the "vision" of a *nation-state* and its outcomes. But in adjudication, all forms of water sovereignty (local, tribal, state, and nation-state) are defined, contested, and renegotiated during the process.

WATER AND IDENTITY ARE NEVER SIMPLIFIED THROUGH LAW

Adjudication was designed to clarify and simplify the state's control of water and how residents were using water so that these private-use rights could be quantified, certified, and mapped. It was about privatizing the use right to water in the state. It was not about commoditizing water per se.[44] Nevertheless, the process revealed what various water sovereigns understood about the value of water, about themselves, and between themselves. Simplification through the state's water-accounting process made water inordinately more complicated, contentious, and capitalized.[45] It also highlighted how identity governance was tied to water.

Like most other western water codes, New Mexico's new 1907 code was designed to award "free water," as long as people made good economic use of it. These western codes served as the water equivalent to the federal 1862 Homestead Act, which put nearly free land into the hands of new pioneer farmers. And just like the Homestead Act, these water codes were never meant to account for prior occupancy, the people already living in the space to be colonized. Indian nations and Hispano acequias in New Mexico preexisted the US colonial-settler state and its new water code policies. The double colonial experience of New Mexico has hardened cultural water governance boundaries. The identity water distinctions do not accurately distinguish the complexity of people. As Juan Estevan Arellano recently wrote describing his family's origins, "We are a mixture of blood from the Iberian Peninsula, Basques and Sephardics and more than likely Moors who mixed here with Mesoamericans, then Pueblos, Apaches, and Navajos, in the case of my kids. All these bloods informed us about how to look at the land and water."[46] States, nation-states, and even Native sovereigns struggle to accept mixed cultural heritage just as states struggle with shared, mixed waters.

In the next two chapters, I turn to the adjudication cases set in the Pojoaque and Taos Valleys, respectively. These cases illustrate the challenges and complexities faced by the adjudicated and the adjudicators since the 1960s. Each case exemplifies complexity for different reasons in different contexts. Aamodt reflects how cultural diversity and legal pluralism in water governance resist simplified state readings of water sovereignty. The Abeyta case in the Taos Valley illustrates how multiple groups of water users came to a negotiated agreement, or water settlement, to preserve local norms of water sovereignty and customary law in use in the valley. Both cases highlight why adjudication failed to "core through" culturally plural views and uses of water. "Litigation illuminates," as one former judge has written, yet it also unsettles relationships between water sovereigns.[47]

2
———

Aamodt, Dammit!

Big Trouble in a Small Basin

Angela and I walked along the dirt road that parallels the often-dry Pojoaque River, a mere trickle moving under a sweltering July sun. As Angela told me:

> I moved here twenty years ago. I moved back, really, after college, because I care about this valley. But my family had no idea about what would happen with this whole [Aamodt] adjudication thing until it was too late . . . By the time they started paying attention again, the deal was done, and it seemed we had already been denied any voice or good deal in that set of discussions. We felt screwed. We still do. We got handed a hornet's nest agreed to by others . . . I'm realistic that the Indians get their water as part of this whole deal. But they can't have everything, can they? And this new deal now forces us to reduce well water use because they [the pueblos] don't like the groundwater pumps. . . . it just seems unfair.[1]

Aamodt, often jokingly referred to as "Aamodt, dammit!" by both former state employees and those who live in the Pojoaque Basin, was the most infamous state adjudication for decades (see map 4). Angela's family had three generations of active defendants in the lawsuit, and she was sharing her latest experiences about the terms of the negotiated agreement. She felt the "non-Indians" in the valley like her had been poorly represented in the meetings that hammered out new water arrangements in the Pojoaque. She remained frustrated, a familiar feeling among her neighbors, too. The settlement had left many in the valley unsettled.

The original Aamodt case was filed by the New Mexico state engineer in 1966, suing almost 2,500 defendants. Centered on the Pojoaque Basin and its small tributaries and the Nambé and the Tesuque Rivers, the Aamodt adjudication suit resurfaced past conflicts and cultural tensions.[2] By the time Aamodt was settled out of court in 2010, the number of defendants had risen to 5,284 and encompassed four

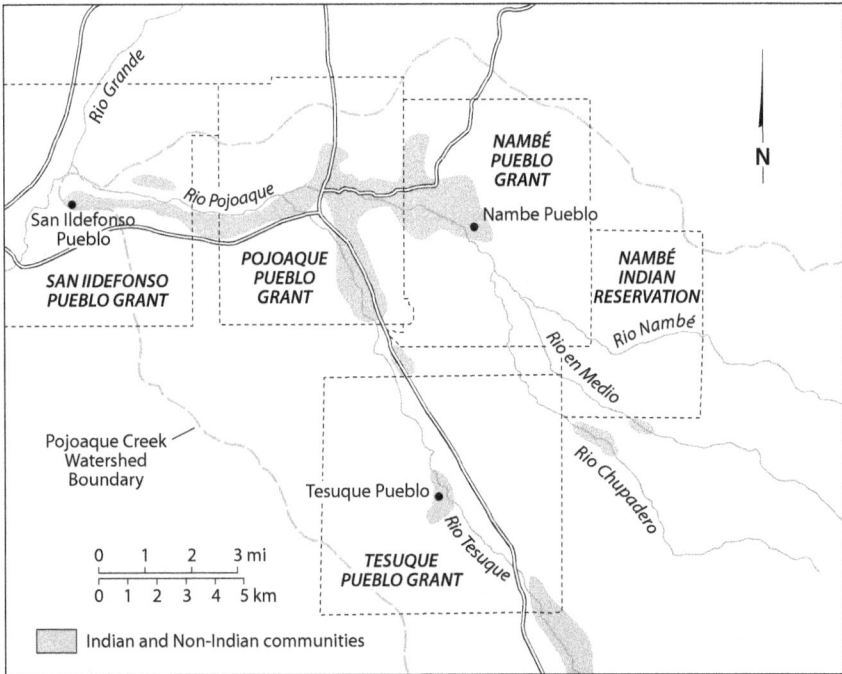

MAP 4. Map of the Aamodt adjudication area, the Pojoaque River Valley, showing the four major pueblos. Adapted from the New Mexico Office of the State Engineer and Utton Center (2013).

Indian pueblos in the Nambé-Pojoaque-Tesuque valleys, as well as a separate irrigation district and some 2,724 acres. Aamodt highlights both the multigenerational complexity of adjudication and the state's ultimate failure to read water across sovereign identity lines. In the end, the four pueblos, federal agencies, and local acequias negotiated their way out of adjudication into a different kind of agreement that was acceptable to the state of New Mexico. Before adjudication was taken out of the courts, however, Aamodt was its own special little hell of a court case.

In this rural commuter valley to the north of Santa Fe, it can seem there are more people and more small land parcels than the limited water can sustain. The Nambé, Tesuque, and Pojoaque are all modest streams for most of the year and often run dry by early July. When they have water, they eventually join the Rio Grande. Water disputes were nothing new in this area. The archives are replete with court records of conflict from the early Spanish Colonial days and through the modern period as residents grappled over the low surface flows.[3] And those are just the cases preserved in official documentation.

With Aamodt opened as a new litigation opportunity, old conflicts over claims to water bubbled up again. Identity questions of Indian and non-Indian finally were confronted and addressed. Non-Indians sought better legal representation and positioning given the strong Indian water rights claims in the valley. Hispanos and Anglos jockeyed for better prior appropriation dates. Groundwater and wells were added late to this process, adding further stress for the valley's residents.

The Aamodt suit produced decades of long, tedious courtroom procedures. Thus, no better case exists in New Mexico adjudications for illustrating the process and lessons from transforming water into a private-use-right property regime. Given the lengthiness of the case, I turned to the legal archives to examine the multiple phases of Aamodt. Some interviewees, like Angela in the opening dialogue, did not remember or know the particularities of the early days of the lawsuit. They often felt trapped by the jumbled legal process that had outlasted generations in their valley. To make sense of Aamodt and its lessons requires time travel and some jumping back and forth through a time line that was inherently messy. The suit was sparked by a project started far upstream.

> It all started, I think . . . with the San Juan-Chama Project stuff, back in the 1960s. That's when [State Engineer Steve] Reynolds started getting serious about adjudication. It was these projects that did it. He realized all this water is going to get connected so better know where people own and use that water, right? I guess we have learned the hard way that once you connect this [Colorado River] water with that [Rio Grande] water, it creates problems, and complications. Everything got more complicated with the connection between the two big rivers. The dams and pipes went in pretty quickly, but the state engineer is still trying to sort out the whole legal thing of what water goes where and who has rights to what part of the two rivers and the Chama River water itself. What a mess! (Tony Adel, Tesuque).[4]

As New Mexico's adjudications accelerated in the mid- to late twentieth century, tied to dam and infrastructure developments, it became clear the process was going to encounter significant hurdles. The Aamodt and Abeyta suits were filed in the late 1960s by then state engineer Steve Reynolds, sparked by the state's need to parse out the water rights involved in the San Juan-Chama Project (see map 5). The San Juan-Chama Project was designed and built to move New Mexico's share of Colorado River water into the Rio Grande Basin. Through a transbasin diversion from the San Juan River in Colorado, a set of pipe transfers into Heron Dam shunts this water downstream to other reservoirs and eventually into the natural stream course of the Chama River. Aamodt was thus complicated in scale from the start.

Such was the reputation of Aamodt that at least two dozen times in other parts of New Mexico, people told me some variation on the following: "At least we're not in the Aamodt case!" Aamodt was infamous and remains so. The most complicating factor wasn't the basin size (it was small) or the limited water. It was legal plu-

Pagosa Springs

160
84

San Juan River
Rio Blanco
Little Navajo
Navajo

COLORADO
NEW MEXICO

Map Area

NEW MEXICO

San Juan-Chama Project

100,000

Chama

Rio Chama
Willow Cr.

N

Heron Reservoir

96,000

Tierra Amarilla

64

El Vado Reservoir
El Vado

Continental Divide

Rio Chama
Cañon de Chama

84

Abiquiu Reservoir

285

Rio Grande

96

Abiquiú

84

68

Española

76

30   285 84

Otowi Gage

1,100,000

Pojoaque

Los Alamos

502

White Rock

8,730

Buckman Direct Diversion Project

White Rock Canyon

Water Treatment

Canyon Road Water Treatment

Santa Fe

4,000

Rio Grande

Cochiti Reservoir

Wastewater Treatment

Santa Fe River

25

285

Legend:

○ Diversion Point
...... Tunnel
▯ Gage
~~ River
◆ Pump station
▭▭▭ Raw Water Pipeline
◀4,000 Approx Average Annual flow (acre feet per year)
← Drinking Water Pipeline/ Delivery Point

0   10   20 mi
0   10   20   30 km

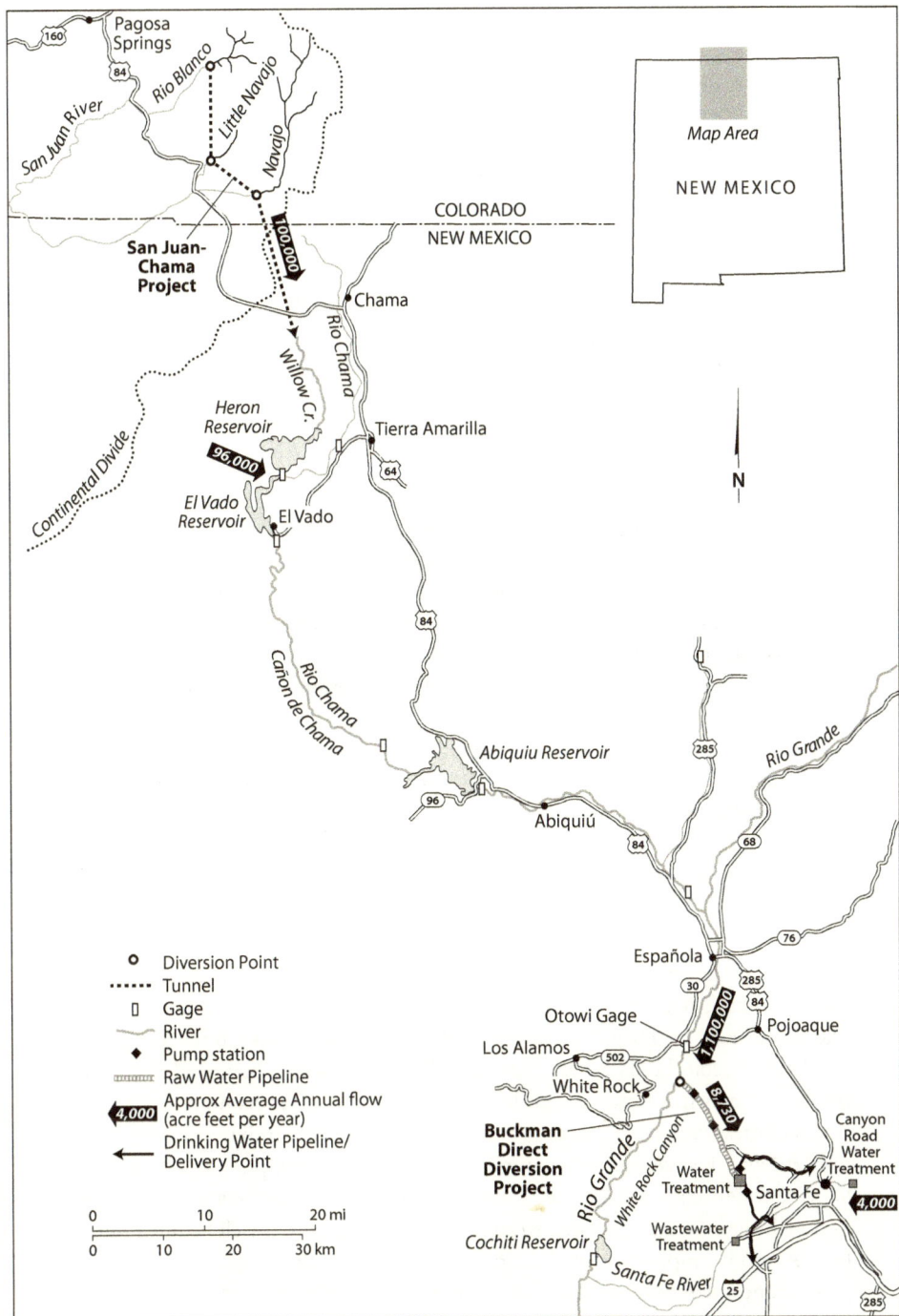

MAP 5. Map of the San Juan-Chama Project. This project was the key infrastructure event triggering the northern adjudications (Aamodt, Abeyta) since the basins would be receiving a share of the San Juan-Chama Project water. Adapted from New Mexico Office of the State Engineer and city of Santa Fe Water Division maps.

ralism. This was a legally complex valley in how residents used water, discussed water, or understood water. The legacy of legal pluralism is reflected in the mountains of legal archival files—so much paper, in fact, that legal scholars refer to the case as two discrete episodes: Aamodt 1 (1966–1984) and Aamodt 2 (1985–2000). Below, I briefly summarize aspects of these two distinct episodes that each lasted more than a generation.[5] Cultural identity, Indian water rights, competing histories of use, inter- and intraethnic disputes over attorneys, acequias and customary water rights, thousands of defendants, ground and surface waters, federal agencies, consulting engineers, anthropologists, historians, and state engineer technicians and lawyers: they were all on display during Aamodt.

THE STRUGGLE TO DEFINE INDIAN AND NON-INDIAN WATERS

The various parties in the Aamodt case—Nambé, Pojoaque, Tesuque, and San Ildefonso Pueblos as well as nonpueblo water users—disputed and contested each other's rights to the water—and not only the waters from the Nambé-Pojoaque-Tesuque (NPT) stream system. The most complicated factor was how Indian water rights would be historically defined and quantified. A long-ignored Supreme Court decision, the Winters case (1908) ruled that implicit water rights existed to support the reserved land base of Indian reservations. The courts never quantified how much water that might be. For federally recognized tribes, the implicit threat of claiming Winters water rights has been a useful tool in negotiating settlements during the last thirty years. The Aamodt case was seen as an opportunity for the Pueblo to address a historical injustice by finally quantifying their water rights. As a long-time attorney for one of the pueblos said, "The settlement process is a bit of a final recourse; since justice was long delayed in getting the Pueblo their proper water rights acknowledged . . . it's simply long, long overdue."[6]

Archaeology and Spanish Colonial archival accounts demonstrate that the Pueblo Indians had long practiced floodwater farming.[7] More permanent canals, like those used by Hispano settlers in New Mexico, are also now common on nearly all Pueblo lands across the state. The pueblos also have real and unmet needs to supply freshwater to residences, casinos, and other economic and recreation facilities. Previous failures to recognize and quantify Indian water rights not only delayed justice but also complicated the state's attempts to document and allocate waters, as I will discuss. Later Hispano settlers, now treated in the courts as non-Indians, were stuck in a strange neocolonial position well before adjudication. Aamodt simply excavated the complex history of their water arrangements.

Hispano *querencia,* or sense of place, was gradually formed in communities over centuries. Early Hispanos settlers were accompanied by Tlaxcalan Native peoples from Central Mexico, and multiple generations of Genízaros (Christianized and converted Plains Indians peoples) were also recruited or enslaved into what became Nuevo México. Families and bloodlines mingled. As a long-time resident of the valley put it, "We're always treated like second-class citizens here even though basically we are genetically the same as the Indians . . . we just don't necessarily claim to be Indian, and they do, so it's complicated, and Aamodt just put a bright flashlight on all this blood politics you know, it's always been awkward when the water issues come up."[8] Hispanos claimed a kind of *settler indigeneity,* as anthropologist James Blair has coined the concept in a separate colonial context, justified by their historical long-term occupancy in the valley.[9] Yet Hispanos remain settlers and non-Indians, not indigenous, in the American juridical context. Parsing through Indian and non-Indian waters would be the first major task for the courts in Aamodt.

As mentioned earlier, the 1952 McCarran Act allowed states to enjoin *federal reserve* waters (including tribal rights) to specific state adjudication practices. With this federal legislation, western states could include the determination of Native water allocations in state adjudication court proceedings and basin research. The Pueblo did not view the McCarran legislation as a positive step. Tensions have long existed between individual states and Native sovereign nations, and the Pueblo were reluctant to acknowledge any power by New Mexico state courts or the New Mexico state engineer on Indian water rights issues. The sovereign tribes and the OSE still retain a degree of legal distance on water issues. To this day, for example, the Pueblo are not required to report their water uses to the OSE.

The legal delays to enjoin the pueblos in state court did not hold up the first technical phase of adjudication by the state. By the late 1950s, surveyors were already at work in the Pojoaque, and most of the mapping work for non-Indian water diversions and water uses was done in less than five years. Between 1966 and the late 1970s, the parties and defendants involved organized their files and strategies. By the mid-1970s, key court cases allowed western states to begin including the Native nations like the Pueblo into state adjudication processes. The courtroom drama of Aamodt then began in earnest, and the pueblos' first legal volley was a massive one. The pueblos contended that not only were their indigenous claims "prior and paramount" in time but that they had senior rights to nearly all the available water in the basin. One can imagine the panic of other non-Indian residents, some of whom had family roots dating back centuries.

The active Aamodt 1 legal phase of the adjudication lasted between 1969 and 1985. During this time, the four pueblos appeared in court to establish their claims to water rights. They were primarily interested in having separate representation for each Pueblo group—four standing attorneys instead of singular Department of Justice representation. Aamodt 1 also determined that federal water protections for the Pueblo would be established. After determining which court would preside and have standing in the matter, things got interesting.

In 1985, the Aamodt II court struck a blow against Pueblo claims to the majority of the valley's surface water, marking the start of the Aamodt II phase of adjudication. Using the so-called Mechem Doctrine, Pueblo rights were to be limited to *historical beneficial use* under the laws of Spain and Mexico. Furthermore, the Aamodt II decision restricted the basis for establishing acreage attached to Indian water rights in the Pojoaque. The court defined Pueblo priority rights based on the acreage irrigated between 1846 and 1924.[10] Aamodt was the first case to use the historical irrigation acreage (HIA) standard as a basis for quantifying Indian water rights.

To explain this seemingly arbitrary range of dates, first recall the historical complexity with the case of the two brothers in the previous chapter. The Pueblo were transferred into the United States as Mexican citizens under the Treaty of Guadalupe Hidalgo in 1848. Following Mechem's (the Aamodt 1 judge) principles, the court ruled that their resource claims would be tied to their date of transfer to the United States as Mexican citizens. Thus, the Pueblo could claim only their *provable* historical water use from the start of the Mexican War in 1846 (when the United States acquired the Southwest) until 1924, when they were redefined as "Indians," and the Pueblo Lands Board was created to supposedly compensate the Pueblo for lost water.[11] In this way, both Pueblo water and identity were tied to Mexican citizenship transfer. Furthermore, their claims were limited to the original Pueblo land grants, as given by the Spanish and Mexican governments, and they could not claim more than the maximum historical planted area.

The Aamodt II court's use of historical acreage for deciding Pueblo water rights was a decisive moment for limiting the Native claims to water in the Pojoaque. Peoples recognized by the United States as Indian (as opposed to Mexican) from their inception were due reservation water rights, falling under Winters Doctrine law. This was a striking paradox: in 1846 the Pueblo were considered "civilized enough," more closely resembling Mexican citizens, until they were redefined as Indians in the 1920s. Because of a strange reassignment of treaty identity as Mexicans once again, the Pueblo did not qualify for a full Winters (Indian) water right treatment. The consequences for future Pueblo water rights cases, if the HIA standards are upheld in future adjudications or settlements, are staggering. The four pueblos, in the end, did not get all of the surface and groundwater in the Pojoaque.

That court decision was celebrated by non-Indians of the valley as a victory since it scaled down the pueblo's previous claims to most of the water. When awarded, Winters Doctrine water rights are more generous than historically calculated figures, which only focus on the maximum extent of past agricultural acreage. Other tribes like the Jicarilla Apache clearly fell under the Winters decision, and the practicably irrigable acreage (PIA; the total land area that *could* be possibly irrigated) standard can hypothetically award more water to the tribes. The distinction may seem a fine one, but it can make a big difference when water quantities are owed to a tribal entity. Ironically, then, tribal sovereigns with little record of sedentary agriculture (such as the Jicarilla Apache) can hypothetically be awarded more water under PIA and Winters standards than the more sedentary Pueblo tribes who were clearly farming for centuries along the Rio Grande. As one past attorney for a Pueblo sovereign put it, rather morosely, "They [the four pueblos here] would have been better off if they had been more nomadic tribes."[12]

Without the historical acreage standard imposed on the Pueblo as former Mexican citizens, the water outlook for non-Indians in the Pojoaque Valley would have been bleak. Paul, a past farmer (now retired) in a small irrigation district in the valley, indicated this to me over coffee one morning in November 2010. He touched his index finger to his thumb, waggling them in a "zero" symbol. "We'd have had nothing, *absolutely nothing*. I wouldn't be here today if Winters rights had been fully awarded to the Pueblo in the valley . . . seriously, I'd have zero and would have to move."[13]

Meanwhile, the situation for Pojoaque Pueblo was perhaps more complicated. It had been depopulated for a period in the early twentieth century and only reincorporated as a pueblo in the 1930s. Because of this, during Aamodt proceedings, there was some question as to whether the Pojoaque group had abandoned their land and water claims. "That kind of story gets dangerous, you know," as Carlos, a member of Pojoaque Pueblo, told me in 2011. "I mean, that questioning of whether we are really 'true Indians' is just unfair, annoying . . . The [Pojoaque] Pueblo struggled because of disease and all these unfair [US Indian] policies that stripped us of who we were, what we owned; sent our kids off to boarding schools to become white American kids . . . ridiculous. And that kind of poisonous doubt continues now—they keep criticizing the casino and saying, 'See, they don't farm, they don't need that water.' As if we have to stay farmers or something to be real Indians. Drives me nuts."[14]

Around the same time, the early 1980s, the OSE realized the need for adjudicating surface and groundwater simultaneously. One of the most contentious decisions by the court on groundwater was in 1982, when the ruling judge declared a moratorium on new appropriations from domestic wells, restricting new wells to indoor water use only. Wells dug after 1982 could not make full use of their originally awarded rights, especially for lawns or gardens outdoors. While this restrictive decision was later modified during settlement, allowing for some light outdoor use in the valley, what did endure was limiting further groundwater appropriations in the Pojoaque Basin. Since the state engineer is not able to restrict domestic well permits when applied for, the court had to do the unpopular work of capping groundwater well development. This decision was rife with controversy.

In the words of Pojoaque Valley Water Users Association (PVWUA) board member Bill Anderson, "In some ways, that was when people got charged up, this whole thing about the wells. . . . I mean, there's no city pipe out there, so people had their own domestic wells, so of course it wasn't about ag[riculture] or even gardens anymore. People in the valley just thought, 'Shit, now the Pueblos are trying to completely kick everyone out of the valley with a well.' So yeah, it got tense."[15]

As tensions between groups continued, the state engineer was steadily determining priority dates for non-Indian water rights holders. By 1982 some twelve hundred non-Indian water users had been sent offers of judgment regarding their water rights, setting up the possibility of new conflicts, as neighbors could contest each other's rights during the inter se process, and individuals could question the state's dates and data.

The period between 1978 and 1987, between Aamodt 1 and 2, marked an active period of litigation, as well as one in which groups formed in an attempt to balance legal interests and representation in the court. In addition, the legal fees for non-Indian defendants were rising, with no way to affordably pay for them. Much of the written record captures the frustrations of non-Indian residents who complained that they could not afford attorneys and that the Indians were getting free government legal counsel from the federal agencies.[16] Non-Pueblos in the valley felt excluded from discussions regarding the quantification of Indian water rights, thus an irrigation district in the Pojoaque Valley reinvented itself as the PVWUA. One of the group's first concerns was the sheer scale of the Pueblo Indian water assertions. Their second concern and priority was to garner funding for legal representation of their interests.

Non-Indian irrigators and property owners lobbied Governor Tony Anaya and even sent letters to President Ronald Reagan to try to garner legal and financial support. They worked on their congressional representatives and senators as well. Congressional representatives from New Mexico coordinated a legal aid fund for non-Indian Aamodt adjudication legal support in the amount of $450,000. While this sum was viewed as small, compared to the federal resources expended for the four Indian pueblos, this appropriation, pushed through by Senator Pete Domenici, was quite a coup for the Pojoaque Valley residents. However, the fund turned out to be less far reaching than they had originally hoped. Much of this was due to internal and external conflicts within and between parties.

One of the defendants in the suit, Doug Martin, remembered that period well and how the early good news on funding turned sour when they realized how expensive representation was for the court procedures and legal proceedings. "It was truly a mess, Eric. I mean we thought we had it made, with the funds necessary to defend ourselves at the beginning. But the process was so long, so drawn out, we could see the funds disappearing before our eyes . . . I served as treasurer, and I could not believe the amount of money our lawyers were billing for each and every thing. We just underestimated how much it would cost, the time it would take. It created a lot of fights in the valley, too, because some of us wanted to stick with particular attorneys."[17]

Doug and his neighbors grappled with the complexity of legal representation and how the federal funds were expended by their lawyers. Furthermore, while the pueblos and their attorneys could make bulk claims for tribal entities, non-Indians had to argue for themselves individually. "We could feel the divide-and-conquer tactics strategy, so we tried to organize differently into a water-user group," Doug said, referring to a new legal strategy born in that period.

QUESTIONS OF LEGAL REPRESENTATION AND INTERESTS

While the fund was useful for court proceedings between 1982 and 1985, serious problems erupted between the non-Indian claimants and their legal counsel over that brief period. The expenses claimed by the legal team led by attorneys Peter Shoenfeld and Larry White were questioned by the PVWUA leadership.[18] The leaders of this nonprofit water-users group were concerned that the legal fund was being spent out too quickly, given the numerous tasks remaining in the adjudication suit. In a series of tense memos and letters and later legal suits and affidavits from late 1984 through 1985, the board of the PVWUA decided to switch legal counsel representation. But the process was bumpy and hostility was barely veiled in the correspondence.

For example, in 1986, the PVWUA sent a letter to all defendants in the case suggesting they sign over legal representation choice to the board members, as they pushed for new legal counsel. Attorney Peter Shoenfeld responded within two weeks to the group's notion to drop him as legal counsel for some of the five hundred defendants in the case. In his letter to his still-then clients, he opined that

> contrary to the material enclosed with the July 25, 1985 letter, you need not do anything by way of response to the PVWUAI, unless you wish to join it. I recommend against joining it. If I do not hear directly from you, you will continue to be my client, and I will continue to represent you in the Aamodt case. My fees will be billed to the federal fund. The PVWUAI is asking you to give it the right to make decisions for you about your water rights. If you join it you will be giving away some of your legal rights. In some documents it asks you to "assign your legal rights" to the association. I suggest to you anyone who does so is inviting the loss of their water rights. The request is reminiscent of the notorious 1880s land grabs in which blank powers of attorney were signed by landowners who soon found out that the Tierra Amarilla land grant, for example, no longer belonged to them.

In this letter, Shoenfeld was opposing the "everyone with 1848 water rights" position (based on the date of the Treaty of Guadalupe Hidalgo) that attorney Marc Sheridan seemed to champion in 1985. Sheridan's position was to get a single priority date awarded to all the ditches, arguing that non-Indians were protected in 1848 as Mexican citizens. This maneuver was meant to avoid later prior appropriation law by adopting a single date, 1848, for everyone who was non-Indian. In contrast, Shoenfeld advocated for individual water rights with differentiated dates. Arguing for a group date was pointless, he contended, as people could be giving up senior water rights (older, high-value dates): "If you detect a note of bitterness in the fore-going, you may be correct," Shoenfeld wrote. He continued:

> You will recall that during our darkest hours, when it appeared that no help was in the offing, Neil, Larry, and I carried much of the financial burden of this case ourselves, purely on trust. For a small group now to decide for you to switch law firms, and thereby to adopt a legal position adverse to the one we so carefully paved over the last five years, is a breach of the trust and mutual confidence we share. I believe the trust will prevail and see us successfully through this lawsuit. I will be honored to continue to represent you if that is your wish. Very truly yours, Peter B. Shoenfeld, signed/printed.[19]

The single letter provides a fascinating display of the high legal and financial stakes embedded in the Aamodt case, indeed in all adjudications. In the first instance, he rhetorically discounts the stand-in representation of individuals and their water rights by a user group (PVWUA). He then cites the historical wounds of lost land grants (such as the Tierra Amarilla land grant in northwestern New Mexico) as a way to reach people for whom this history lives on and remains painful, connecting past land adjudication with then-current water adjudications. Additionally, he challenged the legal merits of the Sheridan plan for representing all valley residents as "former Mexicans" with a single priority date of 1848 to share the water together as a community. Finally, Shoenfeld ends with a plea about his special role in representing the community when there was no legal fund. While this level of detailed correspondence may seem unusual, it is in no way exceptional to this case or others.

The case bumped forward as the attorneys jockeyed for position. In 1986 legal briefs on priority dates for non-Indian water rights holders were requested. The following year, the courts gave hope to the non-Indian irrigators, ruling that the four pueblos involved had historically irrigated 841.5 acres of tribal lands, excluding reservation and replacement lands scheduled to be decided by the courts in October. This final accounting was far lower than the 12,000 acres that the pueblos had claimed initially. For the pueblos, the ruling was seen as another injustice, in addition to questions being raised by their variable histories of land occupancy in the basin. Pojoaque and Tesuque were rather late in organizing as new pueblos, in contrast to Nambé and San Ildefonso, which both had continuous records of occupancy. It wasn't until 1993 that the court ruled that the Nambé and San Ildefonso Pueblos had Aboriginal water rights on reservation lands based on actual, historic use. Even for these two groups, however, a distinction was made based on actual, historical use, and uses that came later in the twentieth century.

Through this archival inspection of legal correspondence, several aspects of the adjudication suit are visible. There was a growing impatience and revulsion about the process among all involved and the feeling that the only people profiting from adjudication were the attorneys. In one of the notes, for example, PVWUA board members suggested that their own lawyers were treating the appropriated court-controlled money as "a legal slush fund" for their own profit.[20] For valley residents, the congressionally funded account was vital, but the finances were drained too quickly.

As one of the former PVWUA board members recalling the 1980s and 1990s in a conversation with me in 2011 said, "It was pretty difficult, but we felt compelled to try and represent the best interests of people for the long haul, and that account was getting tapped pretty quickly by our attorney through his billable hours. We just estimated that we could save money and shorten the process . . . [laughs, drinks coffee] . . . guess that shows you how much we knew what . . . almost thirty years ago? And they [the Aamodt parties] are just now settling and funding this mess. What a charade."[21]

In the back-and-forth correspondence between the PVWUA and legal counsel, it helps to remember that a palpable change was happening in non-Indian legal strategy. The new legal counsel, Sheridan, favored a valley-wide priority date instead of the strict individual prior appropriation plan his predecessor had advocated. This reflected a shift to cooperation between non-Indian parties instead of trying to line up in priority order during legal procedures. It also reflects what Hispano irrigators had been doing for centuries—largely, ignoring individualistic prior appropriation and simply sharing water in their respective valleys. This strategy of legal aggregation for single dates was risky, given that the state water code called for individual water dates, but was a logical tactic for questioning whether prior appropriation could work in this cultural water context.

Sharing the water, then, returned as a near-term legal strategy and goal for water users in the Pojoaque Valley, reflected in the push for a single priority date on numerous ditches. In most respects, this is not surprising—the PVWUA's leadership board had offered to meet with representatives of the four pueblos back in 1982, without attorneys, to see if some discussion could occur. Sharing was long the norm in this region.[22] These advances by the acequias toward the Pueblo to negotiate or just have a dialogue were rebuffed at the time, although whether that was because of the respective tribal councils or their attorneys is unclear. Sharing may have been seen as a distraction tool since new attention was being paid to the hundreds of domestic wells that exist in the Pojoaque Valley.

ADJUDICATION EXPOSES SOVEREIGN
WATER TENSIONS

The consequences of the suit were painfully clear on the ground. A reporter in the early 1980s recorded the fears of couples who had intermarried between Hispano and Pueblo families: "Please don't use my name," a man begged the reporter. "We just rent this land and it belongs to the [Pojoaque] pueblo, and I'm afraid they'll evict us. But I think that all people should have water rights. It's a God-given thing. People here have shared the water for hundreds of years." His wife nodded in the background. "I just don't understand it," she said. "The Indians have always gotten along well with the Spanish here." The numerous additional accounts of inter- and intrafamily difficulties caused by the Aamodt case in the 1980s make for fascinating, if difficult, reading. Many at the time hoped they could settle "by law instead of by guns or fists," as one young man married to a Pueblo woman told the same reporter. The reporter summed up the situation as such: "One thing that nearly everyone in the Pojoaque Valley agrees on is that the divisiveness caused by the water-rights suit has been a shame."[23]

Much of that early social disruption generated from the lawsuit remains as Murphy Inerque, a Pojoaque Pueblo resident and a "veteran of the case" (as he put it), shared with me one abnormally warm November day in 2010:

> What it generated was a lot—I mean a lot—of resentment when the Pueblos tried to move on the entire basin to claim it was all their water. Sure, we live on areas of the valley where maybe they farmed in the past but to try and cheat all their current neighbors . . . Well, people took it badly. I lost a dozen or so friends because of that dispute, or the lawsuit . . . and people just don't forget. Even if the thing [Aamodt] has been settled now, talk to anybody in the valley who is not an Indian and you'll get some hard stares. That deal continues to cut in the valley, not in a good way.[24]

The friction sparked in the late 1960s smoldered for decades, even as the case went through active and passive periods of court litigation, most of it unseen in briefs, claims, and delayed court hearings. A long-time resident named Orlando, when I brought up the tensions in the valley, told me:

> I mean, we know that the Indians were here first. But we've also been here now for some three hundred years, so it was disappointing that we couldn't just use the same water-sharing principles that we had always used together. A lot of us who mobilized into the Pojoaque Valley Water Users Association were just concerned that they would get all the water, all of it. So we understood the priority rules in the end. . . . just that now [2011] it's going to be hard for people to reduce their groundwater use from the well, especially those folks who came after the mid-1980s. That's where the pinch is, so it's still first here, you win; you move here late, tough, you'll have less water on hand.[25]

The tension in the valley mounted with the inclusion of groundwater wells. This was no longer just a visible water conflict, based on surface streams. Changes to relationships in the valley were driven by the visible and the invisible waters being claimed. Given these challenges, it is understandable how and why Aamodt was the longest-standing adjudication suit. Its infamous reputation in New Mexico was well founded. The case was already forty years old when I started work on this project in 2006, older than I was at the time. How can a case last this long, I asked, still a few years into my research. "Ask the lawyers," was the answer from Ernesto, a Pueblo irrigator from Ohkay Owingeh (ex-San Juan Pueblo). He lived outside the basin and had nothing at stake in Aamodt. He had approached me after a presentation at the University of New Mexico, where I had shared some of this work. He grinned. "You'll never finish, man. It [adjudication] will chew you up and spit you out just like it does the rest of us." He did get serious, losing his smile. "And they haven't even bothered to talk to my pueblo yet—it's going to take forever, I think."[26] He was still shaking his head after I thanked him for his questions and remarks.

Ernesto knew that Aamodt had dragged on for more pertinent reasons than just the lawyers and their squabbles. However, his joke provides insight. In addition to bringing old cultural conflicts back to the surface, adjudication also sparked antagonism between professionals. Many attorneys have made a living from this single case. No wonder, then, that even claimants had problems switching attorneys; legal counsels knew a good thing when they saw it. The correspondence between the PVWUA, Shoenfeld, and Sheridan make it clear how tense and interested parties were to retain a role in the suit. The valley residents were not the only ones pitted against each other. Even the professionals were at odds, and the court files and transcripts have elements of a soap opera, just with more technical language. Beyond the definitions of Indian and non-Indian water rights, Aamodt (dammit) left its mark on valley residents and relationships.

The reverberations from the Aamodt adjudication rippled into villages, upstream and downstream, for decades. Even among the valley's acequias, the question of whether to push for individual water rights by specific dates heightened tensions along the ditches, between ditches, and between communities. It unsettled long-standing notions of water sharing throughout the Pojoaque Basin. Proving historical "priority" created skirmishes between communities that had in the past shared waters either informally or through *convenios* (accord or agreement) that established the splitting of water in dry basins.

To illustrate that these struggles were not just along or across perceived ethnic lines (Indian, non-Indian), I turn to an example of a prior convenio and a small stream system in the upper watershed of the Aamodt adjudication area, the Rio en Medio (refer back to map 4). Two tiny villages have shared the perennial Rio en Medio stream for over a century. Aamodt resurfaced the fights over prior appropriation dates at first, but the villages eventually reaffirmed the spirit and the letter of their old accord.

CHUPADERO: WATER SHARING, DATING MADNESS, AND DITCH ACCORDS

The two villages of Rio en Medio and Chupadero, located just north of Santa Fe, have long shared water from a splitter box, a simple concrete device that parts waters on the Rio en Medio stream and diverts a bit more than half to the village of the same name. The lesser amount of flow goes into a connecting canal leading to the upper acequias of the Chupadero basin. Chupadero gets its name from the ephemeral nature of the stream, which is sucked (*chupar*) into its bed and disappears to the naked eye along sections. Looking at this small village, nestled in a verdant strip and surrounded by the bone-dry foothills, gives additional perspective on how adjudication and prior appropriation raise concerns in areas without cleaved identity questions.

Chupadero was first settled in the 1860s. Like most places in northern New Mexico, it was sparsely settled until the late 1800s, when New Mexico was still a US territory. It's still sparsely settled, with only 362 residents counted by the 2010 census (Rio en Medio is even smaller, with 131 residents). The Rio Chupadero runs through the heart of its namesake village, usually as nothing more than a wetland trickle during irrigation season but providing a green ribbon of life in this dry, stark, and striking landscape. The valleys here are deeply dissected and can seem like worlds unto themselves. However, they were never completely disconnected or isolated.

By the late 1800s, it was clear that there was not enough water in the Rio Chupadero to provide for year-round cropping. In 1897, residents created a written convenio with their neighbors just to the north in the village of Rio en Medio to share the water that came from the river by the same name. That agreement details a water-sharing plan whereby "more than half" of the water at the splitter box is to go on its normal path to the village of Rio en Medio, with the right half of the box's flow going on to the ephemerally dry Rio Chupadero channel. A connection canal was built after 1897, between the splitter box and the actual dry channel of the Rio Chupadero, to ensure flows reach the upper ditches and the village of Chupadero.

To this day, one can visit the 1897 splitter box on the Rio en Medio. The right-hand side diversion carries a portion of the flow down a transition canal, cut along the contour and gradient, and then releases it into the natural channel of the Rio Chupadero (see figure 3). This flow feeds into the acequias and several ditches, one of which goes down to the village of Chupadero. It is an intricate, elaborate, gravity-fed system like most of these hand-dug ditches. It is also an example of a transbasin diversion that predates most of the massive twentieth-century federal projects.

The arrangement is an example of what was "normal" yet informal between acequia villages throughout New Mexico. The fact that the two villages were compelled to put their agreement in writing and file it with the Santa Fe County Clerk is evidence that New Mexicans were already aware of the new legal culture slowly remapping their state. Most agreements on water sharing were largely done with a handshake, or standing oral agreement, prior to the late nineteenth century. Forcing prior appropriation onto the acequia villages and stream systems created

FIGURE 3. A mayordomo stands on the banks of the connecting canal that brings water from the Rio en Medio splitter box, agreed to in the 1897 convenio, to the upper ditches of the Chupadero Valley. Photo by the author, 2009.

antagonism, even in tiny villages with a long history of sharing. With adjudication, water was no longer a shared, communal enterprise. It was state property yet with property-use rights determined by historical dates, and the state allocated water according to what people were using on their land.

This legal change started to rework the connections between people and water, relationships between water users, and even worldviews on the purpose of water. The convenio was a legal performance for the territorial courts prior to New Mexico becoming a state (1912), but it officially inscribed the two communities' long-held beliefs about shared water. In legal language and in writing, the convenio formalized the informal to make it legible for the state. Here, too, infrastructure and customary law were tied together. Yet the convenio did not stop all challenges, the most notable of which were revived by the Aamodt adjudication case. Water rights are ranked by first beneficial use date. Under prior appropriation, water users strive to prove the earliest use date possible, hoping to secure senior rights.

Aamodt spurred a scramble for earliest use dates among residents of tiny Chupadero and Rio en Medio. In some cases they contested the dates found by the contract historian and dates in their own oral histories. The OSE contract historian at the time, John Q. Baxter, determined a first date of stream diversion and beneficial use of 1878 for Rio en Medio, which suited those in Rio en Medio just fine. Some upper-ditch users in the Chupadero Valley claimed and were awarded an earlier date of 1863 for the natural flow of the Chupadero River only. The bulk of the flow, however, for both valleys, is dependent on the original 1878 diversion from the Rio en Medio. Only the upper ditch of the Chupadero, then, can claim some natural Chupadero River flow as a prior date. But not much water flows in this stream system without the augmented Rio en Medio waters channeled from the canal (in figure 3).

Hence, a single date of 1878 generally governs the entire Chupadero stream system, save for that one upper ditch. After several decades of squabbling over what the exact priority dates would be in each valley and parcel, they were bound back to the agreement they had in place since 1897. They had to abide by the convenio (see figure 4) and the historical record accepted by the state.

The two villages had none of the legal identity binary fights present in the wider Aamodt case. They had long shared the water from the Rio en Medio. However, Aamodt brought up both old arrangements—and debates about phrasing and meaning—as well as new temptations to leap-frog individual priority dates. Residents of both villages were lured into claiming more "senior" dates to ensure first-in-time access to water but ultimately fell back to the 1897 arrangement. The story of Chupadero and Rio en Medio reflects the struggles of a communal water culture being forced into a system of individual water rights based on prior appropriation. In the end, adjudication did not change much, other than raising hackles between the two valleys and between ditches. Their resolution, to continue sharing the limited waters, highlights the potential solutions hidden in local, historic accords that preexisted the state's interests in water rights.

In 2016, to add to the saga, I met a water user from Rio en Medio at the statewide meeting of the acequia ditches who raised the question about reclaiming some of that shared water. He wondered aloud, "if some of that winter water doesn't legally belong back to our village (Rio en Medio) since no one is irrigating [in Chupadero]."[27] He was hoping for water to fill his stream in the nonirrigating season, an

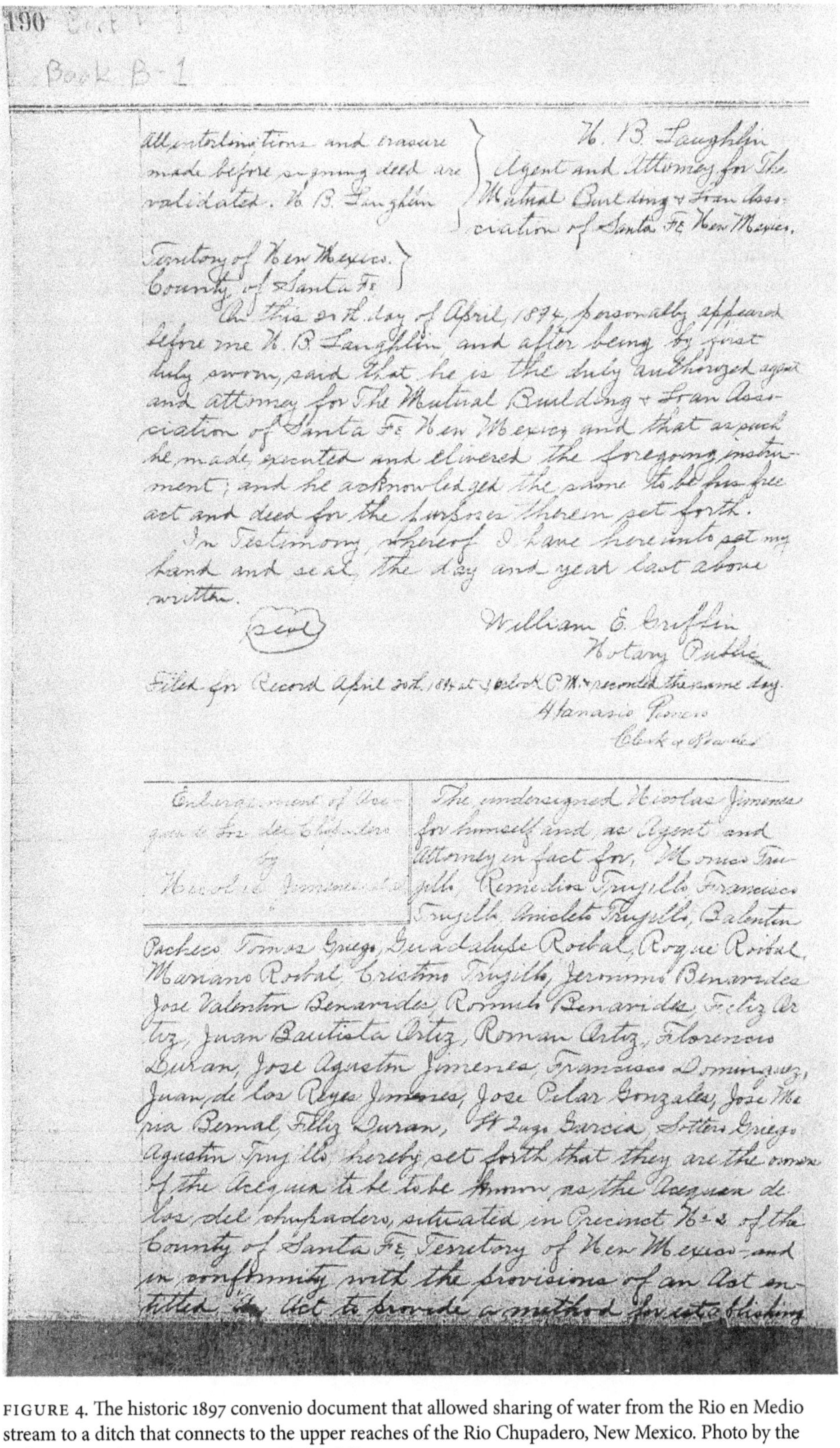

All interlineations and erasures
made before signing deed are
validated. W. B. Laughlin

} W. B. Laughlin
Agent and Attorney for The
Mutual Building & Loan Asso-
ciation of Santa Fe New Mexico.

Territory of New Mexico }
County of Santa Fe }

On this 20th day of April, 1897, personally appeared
before me W. B. Laughlin and after being by first
duly sworn, said that he is the duly authorized agent
and attorney for The Mutual Building & Loan Association of Santa Fe New Mexico, and that as such
he made, executed and delivered the foregoing instrument; and he acknowledged the same to be his free
act and deed for the purposes therein set forth.

In Testimony whereof I have hereunto set my
hand and seal, the day and year last above
written.

(seal)

William E. Griffin
Notary Public

Filed for Record April 20th, 1897 at 3 oclock P.M. & recorded the same day.
Atanacio Romero
Clerk & Recorder

Enlargement of Ace-
quia & for use thereof
by
Nicolas Jimenez etc.

} The undersigned Nicolas Jimenez
for himself and as Agent and
Attorney in fact for, Thomas Tru-
jillo, Remedios Trujillo, Francisco
Trujillo, Anicleto Trujillo, Valentin
Pacheco, Tomas Griego, Guadalupe Roibal, Roque Roibal,
Mariano Roibal, Cristino Trujillo, Jeronimo Benavides,
Jose Valentin Benavides, Romulo Benavides, Feliz Or-
tiz, Juan Bautista Ortiz, Roman Ortiz, Florencio
Duran, Jose Agustin Jimenez, Francisco Dominguez,
Juan de los Reyes Jimenez, Jose Pilar Gonzales, Jose Ma-
ria Bernal, Felis Duran, N. Jose Garcia, Sotero Griego,
Agustin Trujillo, hereby set forth that they are the owners
of the Acequia to be to be known as the Acequia de
los del Chupaderos, situated in Precinct No. 2 of the
County of Santa Fe Territory of New Mexico, and
in conformity with the provisions of an Act entitled an Act to provide a method for establishing

FIGURE 4. The historic 1897 convenio document that allowed sharing of water from the Rio en Medio stream to a ditch that connects to the upper reaches of the Rio Chupadero, New Mexico. Photo by the author. From the Santa Fe County Clerk's Office.

aesthetic and water-for-the-river argument. No action has yet resulted from this, but his idea highlights that new valuations of water—amenity, aesthetic, or ecological in nature—may shift the arrangement again at some point. This short story about two villages that share much in common, not just water, also highlights that adjudication was not just about identity issues in the Pojoaque. Adjudication roiled the customary arrangements and water sharing, even as the larger Aamodt case was reaching settlement phase negotiations around the year 2000.

THE RELUCTANT SETTLING FOR VISIBLE AND INVISIBLE WATER

As Chupadero and Rio en Medio were inching toward reaffirming their standing agreement, so was the larger Aamodt case. A formal settlement process finally began in 2000 and was reached in 2006. By 2010, forty-four years after its initial filing and thirty years after the death of the state engineer (Reynolds) who filed it, the Aamodt settlement was finally and formally funded by the state and federal governments.[28] By then, the case had outlived multiple judges and at least a dozen OSE and private attorneys. Lee Aamodt, the first listed defendant in the case and a scientist from Los Alamos, had become better known for this case than for his scientific contributions. The Aamodt case was its own live, legal reality show before such a thing existed. Settlement would be almost as exhausting as adjudication. But everyone was tired of formal litigation in the courtroom.

As one of the Aamodt settlement instigators told me back in 2010, "I'm honestly tired of it, Eric, and I won't go back [to litigation] no matter how much they hate me." We were discussing the recently funded settlement act. He continued:

> Oddly enough, some of the group had tentatively approached the Pueblos back in 1983, asking for an informal meeting without the lawyers . . . to see if we could come to some discussion points or consensus on what everyone expected out of this. Most folks in the valley . . . maybe you know this already . . . are related, often genetically or through family, or as godparents through the Catholic Church . . . so it made sense to try and create an agreement even back then . . . [shakes head] amazing it took another twenty years for everyone to realize it was time to settle and take it largely out of the courts. Now, there are fewer judges and special masters but just as many attorneys wetting their beaks.[29]

As I write this in 2018, the final settlement has been accepted and stands as the final decree from the court, even if some non-Indian valley residents remain concerned or alarmed about the effects of the terms and the implications for their groundwater well rights. All the parties are still preoccupied with the consequences and costs of a new $261,000,000 regional water system that was one of the negotiation points to make the settlement happen.[30] As I discuss in more detail in chapter 4, this new regional water system will serve the needs of the four pueblos, and the settlement terms encourage non-Indian well owners to cap their wells and hook up to the new regional water system. Groundwater may continue to complicate water rights in the Pojoaque Basin. With the Aamodt settlement, groundwater rights were spatially and historically parsed into new categories of water citizenship based on the dates of well water. The big deal in the Aamodt settlement is this: wells established after 1983 are subject to restrictions and a harsher cap on use.

Around 2300 wells exist in the NPT basin, and some 915 wells were established after 1982. Their active cap on withdrawal was set at 0.7 acre-feet per year (AFY) should well owners accept the terms of the Aamodt settlement. Instead of a standard domestic well award of acre-feet per year, if non-Indian well owners decide to keep their well, the limit to using that quantity is being lowered in the settlement language.[31] This amount is less than the water rights awarded under the OSE groundwater permitting system. Some straws, in other words, were made smaller. Those who sign the settlement and agree to cap groundwater use or tap into the surface waters of the planned regional water system get better terms for continued use of groundwater wells and some degree of relief from future "calls on the basin" water from the nearby pueblos. Those who refuse to sign the settlement and do not hook up to the new regional water system are subject to a hard cap limit (0.3 AFY) for indoor and outdoor uses. The dates, terms, and well restrictions (especially) were so complicated that the simplified "frequently asked questions" document circulated in 2014, during public hearings on the settlement, was still twenty-six pages long. Groundwater remains a third rail of water politics in the valley today.

The finalized settlement still raises the ire of property owners with wells. I spoke with two people running a new nonprofit based in the Pojoaque Valley in July 2015 to learn their concerns about the settlement terms. "It's just so big," Tre Robinson said. "We feel like we're butting heads against something that is too large for us to contest . . . yet they refuse to hear us out. They just want to be done with it and walk away, but we have to live here with the terms."[32] Her friend sighed heavily and picked up where she left off.

> The depressing part is what this settlement has done to us as a community. We lived as neighbors, and even though the Pueblo people think of me as "Hispanic," I'm more Native than Hispanic [by DNA testing, she claimed], so it has really ripped at the seams of our towns and communities in the village. The Pueblo want to be in charge of the rest of us, with no conditions set on how they will run the regional water authority, and we're opposed to that. It's just a handover of the whole valley to the pueblo [San Ildefonso], and we're just really uncomfortable with that."[33]

The implications of the settlement remain unclear to most people who live in the Pojoaque Valley and so is the future impact of yet another water intake (the planned regional water system) in the Rio Grande. Much of what was, is, and remains complicated about the Aamodt case has to do with the sticky, layered notions of cultural identity. Numerous agencies and water jurisdictions were at play here, as the federal government was brought in for the defense of Indian water rights. The four separate pueblo groups remain involved, along with Hispano and Anglo-American signatories to the settlement. The cultural complexity and the layered legal pluralism made for a longer and thornier legal case. Here, diversity complicated the pace, scale, and complexity of adjudication and settlement.

In interviews, it was clear that few well owners in the NPT valley understood the full terms (and historical geography) of the Aamodt case. One recent transplant to the Pojoaque Basin shook her head and said, "It's like they dropped us in the middle of a labyrinth and pretended we all knew how to get out of it." Her confusion is understandable. The settlement took nearly as long as adjudication, and parties who were informed twenty years ago as to what might be "in" the settlement may no longer be those worrying about water in the NPT basin. A whole new set of residents and landowners are trying to make sense of what a capped level of groundwater use will do to their property values. A stage analogy may be trite, but it is accurate: It is the same play, but the entire cast has changed in the last twenty years, and the current actors do not understand the point of the play. The script was handed to them by the previous generation. Those who were in the adjudication "production" twenty years ago are dead or no longer active in their water associations; they want out. It all has a Dickensian quality to it. Even in the latest public hearings in April 2017, Bureau of Reclamation officials tried to calmly explain how the new regional system resulting from settlement would work for new residents just learning about the changes to groundwater and surface waters in the valley.

AAMODT AND THE SUM OF ALL FEARS

Aamodt, as both a former adjudication and a decreed settlement, illustrates three important facets. First, the legal process was slowed by the density of small parcels, the large number of defendants (more than there were acres in the valley), and cultural-legal pluralism. If any suit underscores the penny-wise, pound-foolish process of adjudication, it is certainly Aamodt. Given how little water exists in the Pojoaque Valley, the state, the federal government, and the private water rights owners have all spent an exorbitant amount of financial and human resources on it. Aamodt insiders involved in the legal work in the Pojoaque Basin often shared the following grim perspective: more money was spent adjudicating, and subsequently settling, the valley's water rights than all the land in the basin was worth. This is so logically incomprehensible that it bears restating. More money, per acre, was spent on trying to understand, map, and formalize water in the courts than the land itself was worth. At the time the agreement was forged, signed, and then funded by the federal and state governments (2010), Aamodt was the longest-standing court case in US history, lasting nearly fifty years.

Second, sorting water by legal identity, the binary of Indian and non-Indian water rights, meant the involvement of both state and federal courts. Notions of historical indigenous water uses were at stake in Aamodt. Whether any, or all four, of the Indian pueblos were historically diverting water from their natural stream courses via permanent canals has always been one of the difficult aspects to prove for archaeologists, anthropologists, and historians of the region. No doubt, the Pueblo were diverting directly from flows using floodwater farming and dry-farming techniques. Whether their approach met the criteria for permanent water diversions using perennial or more permanent canals was barely raised, even though this is the basis for state water rights under the 1907 water code in New Mexico. The 1908 Winters decision was an early leverage card for the Pueblo to use to achieve some measure of water justice, but in the end they were treated like Mexican citizens in the transfer to the US system of water rights, getting only historical acreage.

Third, identity in the valley was made more complicated by non-Indian claims to indigeneity. Hispano residents have long claimed that they were allowed to settle in the area near the pueblos, where arable and irrigable land existed centuries ago. Adjudication and consulting historians clearly highlighted that Hispano rhetoric and claims about "sharing water" with the pueblos may have been about Hispano encroachment on Pueblo lands and waters before Spanish arrival.[34] There is no getting around the double history of settler colonialism in this valley and so many others in New Mexico. Hispanos in New Mexico remain legally stuck in the liminal space between indigenous peoples and the Anglo-Americans that arrived after 1848.

The Aamodt case was one of the longest federal court cases in the history of this country. It was also one of the most divisive in New Mexico, pitting neighbor against neighbor and adding binary fuel to the fire of identity. It cleaved water in more defined, cultural ways, separating out Indian versus non-Indian residents. Adjudication in the end treated the Pueblo Indians once again like ex-Mexican citizens. The settlement stemming out of adjudication did not heal these cultural wounds or distinctions. The spark that lit the Aamodt adjudication, the San Juan-Chama Project, continues to have wide-ranging effects on New Mexico's water landscapes, both legal and physical.

San Juan-Chama Project water now goes to various New Mexican cities, including Albuquerque and Santa Fe, for drinking water, as discussed in the following chapters. Building the infrastructure was easier and took less time than the legal adjudication of those same waters. The Navajo Nation got a portion of these project waters (in 2005), as did the Jicarilla Apache in an earlier settlement (1992). Smaller portions of project water were dedicated to Taos (initially the town), and a small allocation was granted to the Pojoaque Valley for the Aamodt settlement in its final form.

Pressure to find an agreement in the Pojoaque Basin between the parties came from another adjudication lawsuit taking place to the north in the Taos Valley. Aamodt was influenced by the Abeyta (Taos) regional adjudication case, and the tangible connection between the two started with, and still depends on water from, the San Juan-Chama Project. The Aamodt parties were influenced and later connected, *socially* and *hydraulically,* by what was happening just to the north of their small basin. Next, I turn to the Taos Valley adjudication procedure (Abeyta) to explain these connections.

3
_____

Abeyta

Taos Struggles, Then Negotiates

Then these guys showed up with survey stuff, walked on the edges of all our fields, and kept talking about a state survey to figure out water use. They had a state vehicle, from the state engineer you know, so a lot of us just decided it was time to meet and start talking. We were doing business with a hand-shake; [it] was a neighborly way of doing business. But it was clear that had to change if they were going to start watching us carefully.

—ENRIQUE MONDRAGÓN[1]

The Taos Valley had its own long-standing adjudication lawsuit stuck in legal mire for decades. The Abeyta suit, filed in 1969, simmered in quiet but adversarial litigation mode as the parties "spent the next 20 years trying to gather and build evidence and find data, maps, and historical documents so that we could annihilate each other in court."[2] In the end, however, the parties did not annihilate each other. Abeyta is notable as the first lawsuit to avoid a "normal" full state adjudication and settle out of court. In its settlement process and terms, Abeyta later influenced the Aamodt suit negotiations to its south, providing a form of interbasin social peer pressure to move to settlement. The lessons of Abeyta have far-reaching implications for basins undergoing and awaiting adjudication.

The major streams in the Taos Valley have their headwaters on Taos Pueblo lands. The Pueblo, then, were in a good position to negotiate their ancestral claims with other parties: the nearby acequias, the town of Taos, and mutual domestic water associations. For the small town of Taos, nestled at seven thousand feet in elevation on the eastern plateau of the Rio Grande Gorge, the strong but unquanti-fied claims to water by the Pueblo were terrifying. The Abeyta adjudication began just as the economy and landscape of Taos were shifting away from agriculture. Like the larger city of Santa Fe to its south, Taos was becoming an art-market mecca, and its economy was increasingly relying on water-thirsty tourism. New residential developments, including suburbs, exurbs, and second homes, also

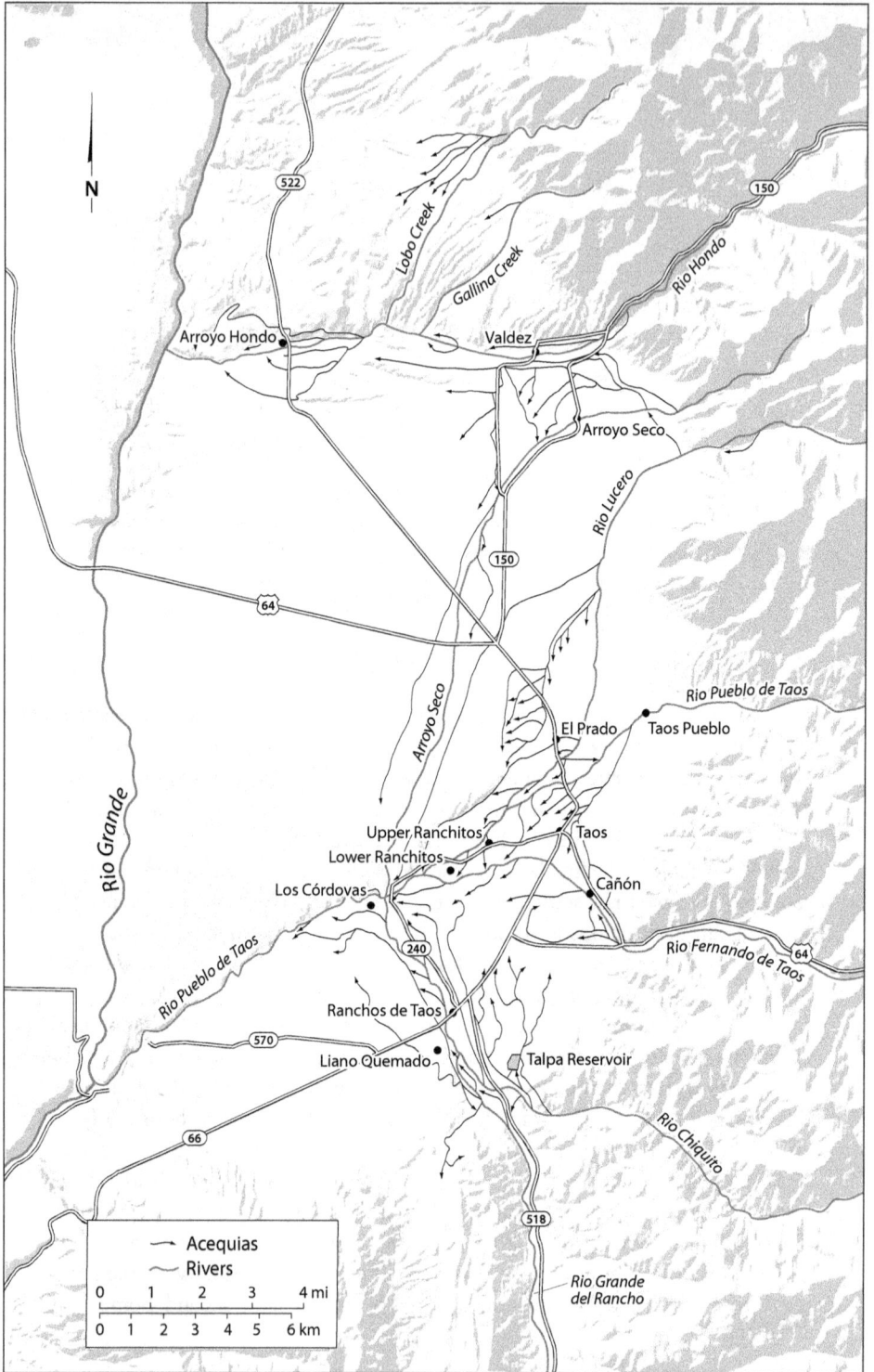

MAP 6. Map of the Taos Valley and its major streams and acequias. Adapted from Rodriguez (2006, plate 1).

demanded more water. With its main water sources—surface stream waters and wells—on the legal table, Taos had a lot to lose in adjudication.

Compared to the Pojoaque Valley to the south, Taos had decent water supplies. Surface water represented nearly 80 percent of water used in the valley, according to a recent water-planning study.[3] Irrigators and landowners relied on a large number of domestic wells, which became a key sticking point in the later adjudication. Water conflict, accommodation, and cooperation were nothing new in the Taos Valley. Long-standing disputes between towns such as Taos, Arroyo Hondo, and Arroyo Seco were par for the course from the eighteenth century onward.

Water sharing between the Taos Pueblo and the adjoining acequias was also tested intermittently yet endured into the late twentieth century. Irrigation in the Taos Valley depends on a set of mountain streams, along with other minor streams to the south (see map 6). These streams feed to one of the densest networks of acequia ditches in New Mexico, some seventy-one individual ditch associations. Approximately two thousand parciantes still depend on these acequias today.[4]

Prior to filing the adjudication suit in the late 1960s, Office of the State Engineer (OSE) technicians had already been at work in the Taos Valley, preparing for water infrastructure. State and federal actions on the long-anticipated San Juan-Chama Project to bring Colorado River water into the Rio Grande Basin spurred the need for state water accounting. Tied to these larger plans was a small dam proposed by the state and the Bureau of Reclamation in the Taos Valley, to be called the Indian Camp Dam. The dam would have been close to where the current and smaller Talpa Reservoir is located (see map 6). While originally popular in Taos in the 1950s, plans for the Indian Camp Dam met with real resistance once federal and state project officials proposed a conservancy district to reorganize water governance. Taoseños worried that the project would raise taxes on already poor farmers. Locals also feared a loss of water governance for the area's acequias and domestic mutual water associations that provide drinking water from wells. The formation of the conservancy district near Albuquerque, decades before, fed these concerns in Taos.

AFTER INDIAN CAMP DAM, THE INEVITABLE

The Indian Camp Dam died because of local opposition to taxation and loss of governance.[5] However, adjudication continued as the process had already been triggered by the OSE and the state's attorney general. After all, a full water accounting was needed for the San Juan-Chama Project and its effects on water rights. The OSE was charged with mapping these lands, waters, and preexisting cultures of waters. It was hard work to envision, much less complete. What is remarkable was the amount of time, precision, and annotation for crops inserted into each map. Each map was made with care yet errors abounded, and the later files are replete with correction maps for boundary sliver issues, ownership changes, and crop annotations that changed over time. Water and water use never stay still, and the maps would have to be constantly updated to be correct.

Mapping, field checking, and aerial photography of all water users and their fields in the valley ensued. OSE field staff came to depend on local irrigator knowledge in Taos. One day in 2011, I spoke with Bob, a retired OSE field technician then in his early seventies. As he recounted his experience in the Taos Valley, what seemed like an expert's account quickly morphed into a humble narrative of long days, confusion, mistakes, and later corrections at the office in Santa Fe. The work was difficult, and Bob's realm of expertise depended on local knowledge to execute it in any satisfactory way, as he related.

> We needed their help to make this happen. I mean, there was a lot of discussion . . . I sometimes felt like they were negotiating with us . . . or that they were trying to maneuver us into decisions that would make it on the map, talking about water duties, or what crops were planted when, bickering about crop rotation and that a fixed amount of water for any field was never set, right? People had a lot to say, and some of these old-timers questioned how or why we were doing all this in the beginning. That field checking must have been . . . in the spring of 1968 or thereabouts. We tried to pay attention to the important stuff on water, but sometimes, well, people just go off, and they got on these tangents how they were special, they ignored the state engineer or the thinking that this wasn't state water, that kind of thing . . . We got it right, mostly, in the end, but a lot of the time we [OSE] were pretty generous about the water duties we assigned or dimensions of fields and how much water people actually needed. I think people were irrigating like crazy with us present [chuckles] . . . I mean they weren't growing rice in places like Taos or Talpa, so that . . . [laughs] was pretty entertaining.[6]

Locals were not always fully trusting of the process. Bob's recounting also makes it clear that Taoseños were intent on being visible to OSE staff in their irrigation practices. They wanted to be seen making full use of their water at the time of mapping, often to the point of overwatering. The hydrographic surveys were the starting point for enjoining the various kinds of expertise necessary to make the 1907 water code work in New Mexico.[7] However, mapping property lines in the valleys was complicated.

Miguel A., a leather shop owner now in his sixties who used to run some cattle in El Prado, recalled local reactions to the new presence:

> When they [technicians from OSE] showed up here in Taos, we kind of freaked out. We didn't really know what they wanted or why they were really interested in field boundaries . . . they started asking a bunch of sensitive questions about water, crops, and stuff. It put us on edge and sometimes . . . you know, we reacted poorly. Some of us were pissed, but those guys were just trying to do their jobs . . . and well we took it out on them sometimes, or refused to talk to them, refused to answer the door, or

ignored all the mail we got [from OSE]. It was a weird time. Maybe the weird part, kind of funny to think about now, is that that [Indian Camp] dam never got built! The whole initial reason for doing it just . . . disappeared. There are some folks who were bitter about that, getting dragged through the courts just to have the whole reason disappear and blow up in the state's face. But at least it [the dam] never got built, so a lot of us were happy.[8]

As the OSE survey continued, there was also growing tension over the needs of a burgeoning small town in the midst of a real estate boom. Even the die-hard agrarians were realists about the needs of the town.

Nestor, one such irrigator from Talpa, a small farming hamlet outside of Taos, was resolutely antidevelopment and antisuburbia. He knew that Taos was trying to preserve its open agricultural space. At the same time, he had no doubts the town would keep growing and so would municipal and residential water needs:

> There's no way to shut down the growth of the town, no way. They will keep the real-tors selling crappy little dry grass plots, allowing domestic water wells, and just keep going until the town reaches the whole foot of the mountains. I've seen it in my life-time. We used to be more like the high villages than what you see today. I mean, thirty years ago [sweeps his hand over the landscape south of town] none of that was there. There was like one shopping place at the edge and then just sage and rabbit-brush and some leftover fields and grazing and stuff. Now look at it. So yeah, the town will get its water, and they'll have to figure out whether they want all the agri-culture still, this green space. I mean, we have fought against that [road] bypass they wanted to put in on the west side, 'cause it was going to cut through some of the best planting land, the *joya* as we call it, prime fields, our best stuff, but they won't be able to have it all ways. Sure wouldn't want to be the mayor of Taos and have to make those tough decisions later on.[9]

For the regional acequias and the Taos Valley Acequia Association (TVAA), their very survival was at stake if the Pueblo won the day on their surface water claims. The acequias were "caught in between" the Pueblo sovereign headwater claims and the residential development happening in the town of Taos. All of the mayordomos and parciantes I interviewed in the valley had examples of acequias struggling with the town. A few acequias had already been impacted by urban culvert demands of new roadways, blocking easy access to ditches. Other irriga-tors were bitter that spring-fed ditches no longer had reliable water because the springs had been dried up by nearby domestic wells. A sense of animosity clouded the valley for well over twenty years.

A PERIOD OF DISTRUST

The period between the early 1970s and the mid-1980s was one of static mistrust, with each set of users fearing potential water losses. Because Taos Pueblo held the headwaters of most streams, there was growing incentive for the other parties, the acequias and the town, to negotiate. Or at least to start the conversation over what had long been a standing agreement over water sharing. The perception by Hispano water users and acequias was expressed by George, a parciante in his eighties who remembered that period well: "Taos Pueblo was going to be in the driver's seat, and at some point, we had to launch discussions with them . . . We just didn't know who to trust or how to trust each other anymore. We had the perception that the Pueblo guys started cutting off our water in the 1970s when it suited them even though we had a handshake understanding that we would share water . . . the state lawsuit just made it tough, and everyone was more guarded."

George shook his head as he remembered that difficult period when the town, ditches, and Pueblo were constantly at odds. "We have to admit now that Palemón [Martinez] was key, as part of the TVAA . . . he was key in getting discussion going. I mean, I don't know if everyone is happy about what is in the [Abeyta] settlement now, but it could have been a lot worse for us. The Pueblo could have tried to take all the water, then where would we be?"[10]

By the late 1980s, as adjudication dragged on, Taos Valley residents were growing increasingly worried. The OSE's mapping, technical, and historical documentary work had been completed by then. "It was done," OSE technician J. Robinson remembered, "but people were just starting to contest what we'd found in terms of water rights, and . . . the fights over other peoples' water rights were just starting . . . yeah, the inter se phase where water users can object to the rights [to water] of their neighbors in the valley. That stuff is what gets ugly in the end."[11] Continued litigation seemed unavoidable until 1989.

People recalled that in the early 1990s, the first rumors were circulating about how the state might use local groundwater resources as Taos' equivalent share of the San Juan-Chama Project surface water award, about 400 AFY (acre-feet per year). The "use" of Chama project water would come not as the Chama water itself but as physical pumping from local aquifers in Taos. This notion of surface equivalency to an increased groundwater use remains contentious to this day. At the time of adjudication and early settlement talks, though, disputes over surface water remained the grist for daily concerns in Taos.

Antonia Gurulé, in her mid-fifties when we spoke in 2010, recalled what such neighborly bickering did to her father. Nicolas, the elder Gurulé, had an especially contentious relationship with two of his neighbors already, and the Abeyta adjudication just made things worse. Living on the border near Arroyo Seco and downstream Arroyo Hondo, Nicolas's small three-acre plot of land was often subject to pitched water rights fights between the two communities. He had to belong to both acequia organizations just to get some water from each, his daughter remembered. After a decade of trying to get corrections done with the OSE for what Nicolas thought were errors in water rights measures, he largely gave up, as she told me:

My father started drinking, hard. I mean he had a drink once in a while, but after that episode . . . he was pretty depressed. He felt like his neighbors abandoned him and let him fall by the wayside. It was pretty sad. Everyone was looking after themselves at that time in the early 1980s and they just weren't being . . . parciantes and compadres anymore. It was just a selfish stage in the whole adjudication mess. It did him in. He didn't lose all his water rights, but he told me he probably lost half of what the family had traditionally used in the past, and he just felt, you know . . . like he let us down or something. He died back in 1991, and still talking about it all.[12]

Nicolas's experience with adjudication highlights what was at stake. This was not just about water itself. Fights for water were personal. When people felt abandoned by their villages, communities, or acequia organizations, the costs were high and deeply wounding.

Finally, leaders of the TVAA, including Palemón Martinez, approached the leaders at Taos Pueblo to propose a settlement of sorts, or at least a dialogue.[13] Serious discussion began between TVAA and Taos Pueblo tribal leadership in 1990 and continued for fifteen years. At first, all concerns were focused on surface water. Over time, groundwater was recruited into the process, as fear that the boom in suburban and residential development and the domestic wells in Taos had been overlooked.

Anthropologist Sylvia Rodriguez, a Taoseña herself, was conducting an ethnographic study of acequias in the mid-1990s in the Taos Valley. Her work underscored these tensions and concerns of the valley's irrigators. Most of her ethnographic work was performed during the decade when the settlement process was being slowly worked out. She noted how adjudication had parsed claims to water through a coarsened binary identity approach. As Rodriguez wrote, "The state underwrites difference by conferring trust status on the Pueblos and other tribes and by ordaining who is or may become a citizen of the nation or province and what that involves . . . Water rights adjudication exploits the difference produced by the Pueblo, the state, the tourist market, and individual actors."[14]

Her findings still echoed in the commentaries peopled shared with me fifteen to twenty years later: "Adjudication divided us. The [Taos Pueblo] Indians were the gatekeepers for the entire watershed." "I cannot imagine this area without flowing water in the ditches." Such concerns remain, it should be noted, even though a settlement was reached.[15] The fundamental divide was the wedge, created by the federal and state recognitions, between valley residents based on identity and differential sovereignty.

New Mexicans paid attention to both of these small basin adjudications in Pojoaque and Taos. Teresa, for example, a water user just north of Truchas (a village to the south of Taos) thought that the bad press surrounding water rights in the Aamodt and Abeyta adjudication suits at least sensitized her and her neighbors when the OSE arrived in the late 1960s to quantify and sort out final water rights in her small valley.

There's only thirty-something people on our ditch . . . I didn't think it would be all that difficult, and we pretty much get along, but we had to pay attention and learn the lessons of the Pojoaque Valley. I mean, we didn't have Pueblo [Indian] water rights here, there's that [nervous chuckle] . . . but seriously, listening and talking to our neighbors to the south in that [Pojoaque] Valley was pretty instructive at the time. We just wanted to avoid the whole mess. There's no reason to spend that kind of crazy state money on a poor valley like this one, with so little water. We didn't have to go to settlement, but we had to pay attention and have our documents ready to go once the state [engineer] started doing work in the valley. And we got finished up with adjudication in the mid-1970s too, so that's good.[16]

New Mexicans living in other areas certainly took notice of the difficulties generated by Aamodt and Abeyta as adjudications.

The state's water-accounting process exacerbated and elevated cultural and identity differences where shared canals and ditches between the Pueblo and the Hispano cut across property lines. This was not isolated to Taos or the Pojoaque Valleys, of course. For all of the historical and contemporary difficulties provoked by adjudication, the Abeyta lawsuit did spark the necessary dialogues between divided water users and those seeking to manage or administer their water rights. If water abounded, such to-the-drop accounting and management might not be necessary. However, by the time adjudications got rolling, water appropriations may already have exceeded actual water availability. As one former high-ranked adjudicator remarked:

I feel like we came at the process decades after there was no more water to go around, it all had a slightly desperate feel and smell to it, like we were just there to look for more water that simply doesn't exist. Statutes demand we do it . . . we ran all over the state trying to find this water, making sure everyone got their due process, but I . . . . just wonder if we couldn't just meter every stream and tributary in the state and have the OSE just pay attention to those areas that need addressing or that go through difficulties. And you know, a lawsuit won't create more water in this state.[17]

However, the OSE, the Bureau of Reclamation, and the politicians continue to propose schemes to augment local waters and to find water from other basins and import it into the needed basin. Such thinking has not disappeared, even if current OSE employees will not quite put it that way.

WATERSHED LESSONS: GEOGRAPHY, ANXIETIES, AND A LEGACY OF SHARING

Abeyta highlighted many of the cultural, political, and hydrologic dimensions trapped in the complicated business of water adjudications. The water amounts involved are minor, but they remain problematic. The advantage that Taos Pueblo held was clear from the beginning: it was upstream of everyone. But this was not just about location. Cultural views on water had to be reconciled. Political horse trading created some of the initial enthusiasm for new water from the San Juan-Chama Project. However, adjudication produced anxieties in the valley. The process changed social relationships between individual water users and neighbors and between whole cultural groups.

People in the Taos Valley learned and relearned their water relationships in some ways because adjudication forced conversations out into the open. John Nichols, who lives in Taos and is the author of *The Milagro Beanfield War*, wrote about how it affected people and changed their perceptions:

> There was a time during the heat of the battle against the conservancy district in Taos that I had tacked up, on every wall in my little adobe house, the State Engineer's hydrographic survey maps of every irrigated piece of land in the Taos Valley. I knew by heart all the acequias, their locations, and many of the pieces of land that they irrigated. I had a big telephone-book-sized list of all the parciantes. I knew many of those parciantes personally. It was the most intimate kind of map you could have of my home area. When I looked at all those parcels of irrigated land and the people who owned them I was learning an entire town. It was like being in medical school and dissecting a body. It was like memorizing the Bible. It was like learning the entire history of a people that had become an important part of my own history and vice-versa. It's one way I put down roots. That's how communities remain strong, when their people have that connection, those roots, that obligation.[18]

Those fears and anxieties heightened over time. Levine, writing about adjudication in 1990, reported seeing bumper stickers proclaiming, *La Agua es Vida; Let's Share it.*[19] By 2016, these messages had become more strident, and I was more likely to spot messages like *El Agua no se vende; el agua se defende* (Water is not to be sold; it is to be defended). Such rallying cries from communal water defenders had to do with the growing resistance to the framing of water as a transferable commodity. The shift in tone over twenty years also highlights a shift in rhetoric, from the old notion of sharing water between the main water sovereigns in Taos, to "defending" water from leaving the area, a message that reflected worries about water transfers out of the valley. Water, once sacred and the basis of village life, had become legally just a "resource" that could be severed and sent away from its basin of origin. These concerns with potential transfers and actual local water rights sales also changed the tone of conversations. Instead of litigating against each other in the valley, Taoseños were on alert about water being traded or transferred away.

In Taos, the issues raised were only slightly less complex than in the Pojoaque. First, only one pueblo (Taos) was involved in the Abeyta case. In contrast, the Aamodt case to the south had four distinct Pueblo reservations, with their own respective legal representation. Second, the TVAA was already an organized and coherent institution, allowing it to play a more effective role in the process. As locals began to undertake direct conversations, it seemed that adjudication as a state process had failed them. The success of the TVAA and Taos Pueblo leadership in bringing parties together to the negotiating table was notable.[20] Sharing the watershed itself would be challenging; the parties knew that. But the town, county, acequias, and pueblo itself were all more worried about water leaving Taos.

As a Taos resident named Alfonso summarized the difficulties of the original process in this valley, "It's [adjudication] just not suited to figure it all out. The [state] water code [of 1907], when they call for this to play out in courts, I mean . . . that was a bunch of Anglo guys and some Nuevomexicanos thinking out loud, but the Indians, I mean they weren't at the table, so who knew back then that the whole part of the code about adjudication wouldn't work for this state, actually most of this state!"[21] This tension of water federalism, the balance between state duties to adjudicate their own respective jurisdictional water rights and the federal trust responsibilities to Indian tribes, remains problematic. This fragmented landscape of water sovereignty and power differentials between parties partially explains why the adversarial aspects of adjudication were poorly suited to settling Native sovereign water claims.

TRANSFORMING BASIN ADJUDICATIONS INTO LOCAL WATER SETTLEMENTS

By the time I first visited Taos in March 2007 to speak with irrigators in the area, the parties to the Abeyta adjudication suit had crafted a draft settlement agreement the year before (2006). Even after a year, Taoseños were nervous to talk about these water rights issues. They remained wary of the terms resulting from the negotiations in the settlement. The settlement was viewed then as a tentative truce. However, the settlement came with the promise of federal funding, direct aid for the Pueblo, and infrastructure projects for all parties involved. Thus, there were financial teeth to the agreement and inherent economic pressures to keep the deal alive into the future. If any party decided to walk away, the consequences could be dramatic financially (for all parties). The final terms of settlements would need to be satisfied through state and especially federal funding. The final result would serve as equivalent to adjudication's "final decree" by a judge. Adjudication started the process of finding waters and water users, predicated on the Indian Camp Dam that was never built. Settlement closed on the deal by forging a new kind of state process. Why did adjudication fail?

Assessing failure or success is difficult. On the one hand, state employees spent years conducting the tedious work of mapping and conducting archival research. The OSE produced a remarkable anatomical map of the valley, its water, and its water users. Thus, the technical phase produced valuable information for the negotiating parties. That cartographic and historical information was used not just by the state but also by people on the ground, leading to a new kind of water-user negotiation. Phase one, the information and data, was hardly the issue. Where adjudication arguably failed was in the adversarial legal process. The acequias, especially under the TVAA, realized there was much to lose. Taos Pueblo leadership realized that adjudication might win them paper water rights, but adjudication itself provides no funding or infrastructure to use those water rights. The parties' desire to short-circuit the full court process underlines the weak points in adjudication: negotiating parties wanted an alternative to the adversarial relationships inherent in the adjudication process. And the traditional litigation route provided no benefits to this long ordeal.

One of the later participants I spoke with back in late 2009 was proudly optimistic they had fairly represented the interests for non-Indians in the Taos Valley.

> We did kind of create the template for these intercultural settlements between the [Taos] pueblo, the town of Taos, and the acequias . . . that all share the same water. That's a big deal, you know? We did this. The state did not. The state engineer wasn't really interested in us settling this case by ourselves. They only came around, I think, because the federal government offered the [Taos] pueblo some money and resources to finally settle all this. I just hope the final agreement goes off without a hitch . . . and that people will follow the settlement rules and use the funding, because some people now feel like they never knew what was in the language of the settlement.[22]

That sense of tempered optimism survives to this day in the valley.

At a recent 2015 public meeting of the legislature's Water and Natural Resources Committee in Taos, for example, parties congratulated each other on the hard-fought agreement. Representatives from both Taos Pueblo and the regional acequia organization recalled the discussions and disputes, mimicking the adjudication process itself, even if there was an understanding they were in "settlement mode." Gil Suazo of Taos Pueblo did remind the crowd that day that the negotiations had stalled in the early 2000s. Not until 2003, when a mediator from out of state, a Californian (Mike Hardy, from Sacramento), arrived was there "any tangible evidence of progress in discussions." The courts, throughout the 1990s, also pushed for clear benchmarks for settlement. As Gil stated for the record, "It helped that the court set a deadline and put our feet to the fire, after all, we could have kept talking for a long, long time, but having to show progress [to the judge] in the discussion was probably the key to resolving it."[23] The parties escaped litigation but could not fully avoid meeting time line benchmarks set by the adjudication courts.

Mr. Cordoba, also from Taos Pueblo, offered another perspective. "The mediator helped, sure," he said, then half-joked, "but we didn't have anything left to throw at each other." His observation was met with nervous laughter in the room. The parties were tired, he said, and ready to make the settlement happen. One of the senators present at this 2015 legislature subcommittee meeting wondered aloud whether[24] more adjudications or settlements could not be taken care of "by a mediator for less money, enclosed in a concrete room with no windows, no breaks, and bad coffee" to speed up the process.[24] More uncomfortable and nervous laughter rippled across the audience at the thought that it might take hostage conditions, and a form of Stockholm syndrome sympathy, to make settlement more efficient in the future.

Social relations and extended dialogue in the Taos Valley helped spark an early impetus to resolve the impasse outside the courts, through settlement, instead of a full adjudication. Often, but not always, this depth of relationship was rooted in kinship. Yet the more diverse a place is, or the more kinship is splintered, the more difficult it can sometimes be to get cooperation. But this sounds too facile, too generic, for what actually happened. As Sylvia Rodriguez made clear in her work, everyone knew what was at stake and what could be lost in the valley if no agreement was forged. Settlement was as much about direct contact between the acequias and the Pueblo as it was about the inherent mistrust in the state's underlying motives. The TVAA saw clear benefits from leaving adjudication, in that water sharing might be arranged with the Pueblo. But all parties saw benefits in the political economy of settlement, something that traditional state adjudications could not offer.

ADAPTING A SETTLEMENT TEMPLATE

As one of the Taos Pueblo negotiators put it, settlements "are a symbiotic process" that serve more parties than the traditional two parties of New Mexico versus defendants in disputed adjudications. Indian water rights settlements are now a preferred route at the federal level for avoiding prolonged courtroom litigation, and New Mexico's experience has reflected this evolution in negotiated agreements. The Taos Valley experience was informed by a settlement to their west, which was also tied to the San Juan-Chama Project and the Jicarilla Apache settlement. The Jicarilla Apache reservation is located near the completed San Juan-Chama Project. Yet their water rights, tied to reservation purposes of development, had never been quantified by the state or the federal government.

The Jicarilla Apache settlement, reached in 1992 and funded by Congress in that same year, set an important precedent for later template agreements between the parties.[25] The resulting settlement awarded the Jicarilla some San Juan-Chama Project waters, including 6,500 AFY that could be leased or sold out of their portfolio of water rights totaling 40,000 AFY. New infrastructure, quantified water, and an ability to create revenue from the project water were major enticements.

As part of that settlement, the Jicarilla agreed to renounce further Winters rights claims, and litigation was ended. The agreement was specific and tailored to the Jicarilla, stipulating, for example, that the agreement would not be "in any way to quantify or otherwise adversely affect the land and water rights, claims, or entitlements to water of the Navajo Nation, or any Indian tribe, pueblo, or community, other than the Jicarilla Apache Tribe."[26] The Jicarilla settlement was the first major Indian water rights accomplishment tied to the vital San Juan-Chama Project, allowing the state, federal government, and other adjudications to proceed with allocating water rights for project waters that would reach the Rio Grande.

Settlement was not necessarily cheaper or shorter than full adjudications or without problems. One of the shortcomings of settlements is that they are isolated agreements, an agreement that may not be based on sound hydrological or legal opinions—it is what the parties agree on, no more and no less. Settlements do not provide useful precedents that can be used in future cases or adjudications. Furthermore, the experience of one Native nation cannot be generalized to represent all others. As one Jicarilla Apache resident and water manager joked during a public discussion on water, "Once you know one Indian tribe, you know *one* Indian tribe." In other words, no tribes are the same, and thus each settlement is different and tailored to sovereign circumstances. And yet, the settlement template became remarkably attractive.

Jicarilla was an important benchmark agreement. For Taos, the Jicarilla template influenced what would happen in the Abeyta case, and Abeyta then inspired Pojoaque (Aamodt), creating a ripple effect across New Mexico. "At the time," quipped D. Ortiz, "we knew the whole Jicarilla deal [settlement] would be useful. It had some language about ending the legal fight that we thought would be useful here in Taos, but we didn't really know how big of a deal it really was in triggering all these lawsuits to try and pursue settlement. You don't really know how important something like that [the Jicarilla settlement] is until you can look backwards, right?"[27] At least a dozen other Taoseños I spoke with, however, thought it was the principle of respectful dialogue and not the "Jicarilla template" that was more important. As one of the principals to negotiating the Taos settlement stated back in 2010, "respectful discussion, listening, understanding needs on all sides, these were the keys."[28]

The Aamodt and Abeyta cases went from being formal adjudications in New Mexico to respective agreements, ending in separate settlement processes. Both morphed into settlements outside the formal adjudication courts because of local interventions. There is much to admire about settlement, in that it does involve aspects of conflict resolution. Settlement forces the parties to converse, discuss, and hash out major differences. It can create goodwill at the table, as parties may agree not to sue each other or not to enforce "seniority" rights on a stream. But they are also lengthy, complex, and expensive in the end.

I am not the first to point out the costs of water settlements for sorting out western water rights. Daniel McCool raised doubts in the late 1980s after the federal government declared its preference for negotiated settlements.[29] He was even

more concerned in 2002, reflecting on nearly twenty years of Indian water settlements: "In hindsight it appears particularly naïve to assume that the most complex, contentious, and long-lasting lawsuits in the land could suddenly be settled
'quickly.'"[30] One wonders if those same Reagan-era officials might walk back their
preference today, given the cost to federal and state taxpayers of these settlements
in western water adjudications. In these cases, the parties will never know if adjudication would have been cheaper or quicker. Nevertheless, attend any hearing
regarding the desirability of adjudication versus settlements when it comes to
Native water claims and you will inevitably hear that "of course settlement is preferable." As someone from Taos informally but astutely joked a decade ago, "Settlement is like adjudication with benefits."

Adjudication is arduous, adversarial, and time-consuming, with little benefit
involved. It also takes more than settlement and dispute resolution to get around
the most antagonistic legal features of adjudication. New Mexicans who felt
ignored by the state, or bullied and outspent by either the state (OSE) or the federal
attorneys negotiating for the Pueblo, needed help with their own legal claims (as
we saw in the last chapter). The resulting settlement ended the longer process of
adjudication yet left out the concerns of many individuals, especially those who
were not part of the agreement process. Notably, public welfare concerns by a Taos
County citizen committee bedeviled the final details of the Taos settlement, especially the components about transferring retired groundwater from Taos downstream to meet Pojoaque's settlement needs. Octogenarian parciante George V.
was a little more philosophical than he was upset:

> Look, we could have done a better job of engaging in all this, but some of it was no
> doubt behind closed doors. That's the problem, right? People object to not knowing
> what happened without their knowledge, but how can we do that? Can it always be
> that everyone is in the room? Can you always trust your mayordomo or regional
> acequia to do the right thing? Probably not, I mean . . . it's difficult, we didn't really
> vote on any of this, and it is problematic especially for those who did not join the
> TVAA. There were like, a dozen or fifteen acequias I think that didn't join the group.
> Now they're upset, I get it. But they didn't join then. Now they want to question the
> whole thing, in hindsight?[31]

So while the potential for future water transfers worries irrigators, Taoseños
clearly saw the benefits of a mutual agreement in the valley.

TAOS VALLEY LESSONS AND KEY
SETTLEMENT COMPONENTS

Abeyta reflects how agreement, negotiation, and final settlements can replace
adjudication when parties enter into a dialogue. In the end, the TVAA was instru-

mental in approaching and negotiating with Taos Pueblo. The Abeyta experience also confirmed the weakness of adjudication as a state-driven process for clarifying water rights when federally recognized tribes have implied or explicit rights to water. To some parties in the suit, Abeyta highlights the failure of adjudication to resolve these issues when water crosses the sovereign identity line into Indian water rights. That relative failure also allowed the local acequias to assert some degree of negotiated room to maneuver in talks with the Pueblo.

As a result of the agreement now in place, the needs of Taos Pueblo, the regional acequias, the city of Taos, and some domestic mutual and sanitation districts were met. Settlement did not solve all problems. Settlement is pricey because of the enrollment of federal and state funding when Indian water rights are at stake. Federal pork barrel spending is not just for dams and water infrastructure; it has also been spent on the legal finery underpinning these agreements. As a result, some $66 million were set aside for Taos Pueblo water rights acquisitions and for water projects, with another $58 million built in to benefit all of the parties in the valley. The state of New Mexico agreed to fund non-Indian water projects and needs to the tune of an additional $20 million under the agreement. Settlement did make strange connections across completely different former adjudications.

As I discuss next, both the Taos and Pojoaque Valleys are tied to each through an awkward retired Taos groundwater transfer to surface water use to satisfy some of the terms of the Aamodt settlement in the Pojoaque Valley. Taos Pueblo was also awarded an additional 2,200 AFY of San Juan-Chama Project water credit, which will be used locally from a combination of ground and surface waters, well beyond the original 400 AFY the Taos area received in the original conception of the San Juan-Chama Project.[32]

What is stunning about the Aamodt and Abeyta cases is how little water was involved overall given the decades of time and resources spent. The quantities bickered over, litigated, and finally settled for pale in comparison to what awaits when the state decides to address the water rights of the Albuquerque and Middle Rio Grande region, as I discuss later in chapter 8. As George V. put it:

> The state spent so much goddamn time and money on us, and it's not much water up here [Taos]. I mean I think people like that we have our water counted now, and it's a done deal, but we're small water, you know? I think since we got stuck with the adjudication and now settlement, seems like Albuquerque grew to twice its size. They spent their time counting our small water beans; meanwhile, the entire field of Albuquerque gets ignored? It's all tiny water stuff up here compared to the Rio Grande stuff . . . they haven't touched it.[33]

The inability of state water adjudications to reconcile and resolve the tangled water, identity, biophysical, and political challenges in two small basins made the settlement process far more attractive. In the end, parties to the suit got their individual water rights but balanced in a way to avoid prior appropriation calls on rivers by Taos Pueblo. The expensive negotiated settlements connected basins and users in completely unexpected ways. Surface and groundwater became mingled both physically and legally. Adjudications and settlements are now increasingly conjunctive in their waters, resolving surface and groundwater, and pose new challenges for settling water disputes in the state.

Local Settlements Connect What State Adjudication Severed

I know the Indians think they own all the water, and that was their strong starting position in the [Pojoaque] Valley, but they need to understand that thousands of us live here in Taos now and that we have to find ways to work together. We can't just travel back in time and pretend that they're the only ones here now . . . we may just be splitting water that doesn't exist from the [San Juan-Chama] project, and all the weird groundwater projects for the local aquifers just don't mesh with how we want to use water. Those new mitigation wells for pumping in the Taos Valley . . . it just feels wrong, like they're hiding the fact there isn't enough water here. The settlement was done behind closed doors, by a bunch of lawyers and feds who didn't ask us about anything. There was maybe one guy [attorney] looking out for us at those final settlement meetings—that's upsetting. It's just not representative, not democratic. I don't get it. Now will the settlement really help us calm down, deal with each other like we used to? I kind of doubt it.

—GERALDINE GURULÉ[1]

Geraldine shared her worries about the Abeyta settlement and its potential consequences. The latest hydraulic arrangements between the basins of Taos and Pojoaque had their roots in the San Juan-Chama Project. Only the Pojoaque Valley has a direct physical connection to the main stem of the Rio Grande that includes the Chama project water. Despite having no physical access to San Juan-Chama water, Taos Valley and its Abeyta settlement are affected by the water accounting and the ensuing water shuffle. As discussed in previous chapters, both the Aamodt and the Abeyta adjudications created or resurfaced old animosities and failed to produce timely results. In both cases the parties went to settlement as an out-of-court legal agreement. Geraldine's fears reflect the unease of many over the murky and seemingly political arrangements that are replumbing the valleys of New Mexico to make this legal settlement work.

The Aamodt and Abeyta settlements connected New Mexican water sovereigns and water in new ways.[2] Negotiated settlements are now preferred by the federal

TABLE 1 Key points and benefits of the Aamodt settlement (Nambé-Pojoaque-Tesuque Valleys)

- Expressly authorizes, ratifies, and confirms the settlement agreement
- Resolves the water rights claims of the Pueblo
- Provides for implementation of a Cost Sharing and System Integration Agreement and an Operating Agreement between the governmental agencies and the pueblos
- Provides that construction costs of the regional water system pertaining to the pueblos are federal costs, which they will not have to reimburse, and that costs pertaining to the county utility are to be covered by state and local entities
- Allocates 1,079 acre-feet of San Juan-Chama contract water for use by the regional water system
- Provides that the pueblos' share of San Juan-Chama costs is nonreimbursable
- Provides $56.4 million in funding now and authorizes an additional $50 million for construction of the regional water system to serve Pueblo and non-Indian residents
- Provides $25.4 million in funding now for the acquisition of water rights and projects to improve existing Pueblo water supply infrastructure
- Authorizes an additional $42.5 million to assist with operation and maintenance of the regional water system
- Allocates over 6,100 acre-feet of water to the pueblos, with various priority dates

SOURCES: NM OSE Settlement Agreement (Aamodt) (2012); Richards (2005).

government to resolve Indian water rights issues, yet the agreements also include benefits for non-Indians. As political scientist Daniel McCool summarized Indian water settlements bluntly fifteen years ago: "It is a matter of politics. Indian settlements are politically feasible only if money is spent on mitigating their impact on Anglos ... Indian settlements have been used in the age-old strategy of the "Indian blanket," which refers to tying a dubious Anglo project to an Indian program in hopes that the added political support will help push an otherwise unacceptable Anglo expenditure through Congress. One example is the San Juan-Chama Project, which simultaneously authorized the Chama Project for Albuquerque and the Navajo Indian Irrigation Project."[3]

These bundling strategies have so far served non-Indian communities in the western states well—in many cases better than they have served the Indian communities themselves, as McCool noted. Whether enough water actually exists to satisfy the new agreements is one major question. Another is how differently some New Mexicans feel about the trade-offs of the settlement, especially when it comes to resolving the thorny issues of groundwater and domestic wells.[4]

In other parts of New Mexico with remaining Indian water rights to address, adjudications will likely become settlements as favored by the federal government and the various Pueblo sovereigns. The legal settlements—triggered by infrastructure projects in these two cases—have paradoxically resulted in *additional* water infrastructure projects that now bind multiple basins. Tables 1 and 2 summarize

TABLE 2 Key points and benefits of the Taos Pueblo Water Rights Settlement Agreement
(former Abeyta adjudication)

- Resolves the water rights claims of the Taos Pueblo and authorizes the Taos Settlement Agreement.
- Allocates 2,621 acre-feet of San Juan-Chama contract water to the Pueblo (2,215) and to other settlement parties (406).
- Provides approximately 12,000 acre-feet per year of total water depletion rights to the Pueblo.
- Provides $66 million in funding and authorizes an additional $58 million for Pueblo and non-Pueblo water development and conservation projects.
- Authorizes federal funding for the planning, design, and construction of water infrastructure projects known as Mutual-Benefit Projects. The federal share of total costs is 75 percent, for which the United States will not be reimbursed. The nonfederal share can include in-kind contributions.
- Accomplishes federal funding through two funds: (1) the Taos Pueblo Infrastructure and Watershed Fund for providing grants to the Pueblo for Mutual-Benefit Projects; and (2) the Taos Pueblo Water Development Fund for the Pueblo's costs for projects such as water rights acquisition, rehabilitation of existing infrastructure, and various watershed protection activities, including buffalo pasture revitalization.
- Authorizes funding for grants to non-Pueblo entities for Mutual-Benefit Projects.

SOURCE: NM OSE Taos Pueblo Water Rights Settlement Agreement (2012).

the key terms of the settlements. In particular, note how both the Aamodt and Abeyta settlements hinge on San Juan-Chama water.

There is much to value in the social terms of the settlements. Aamodt and Abeyta settled indigenous water rights and also stipulated that the pueblos would not put senior water rights "calls on the river," shutting down their neighbors. Abeyta went a step further in implementing an actual water-sharing agreement that was long practiced between Taos Pueblo and the signing acequias that were party to the settlement. The acequia associations that did not join in the settlement, about fifteen of them, are left wondering about their future water allocations since they chose not to participate. Aamodt and its settlement language have no such provision for water sharing. In times of drought, the four pueblos will get their senior water rights fulfilled first.

The most contentious component, by far, from the intertwined Aamodt and Abeyta settlements was the water rights transfer of groundwater from Top of the World Farms to be put to surface water use downstream on the Rio Grande. Located in Taos County, near the Colorado border, the farm owned rights to substantial groundwater, which it used for pivot irrigation. Although in the Rio Grande Basin, the farm is located some ninety miles north of Pojoaque, the focus of the Aamodt adjudication. However, Top of the World's water rights became fundamental to resolving the Aamodt settlement. In accordance with the settlement, 1,752 acre-feet per year of Top of the World groundwater rights were retired from use and then transferred as surface water rights to Santa Fe County and the

Bureau of Indian Affairs. No water will actually be piped across basins. No new water is physically entering the Rio Grande. This water transfer is treated as an equivalent on paper, illustrating the fictions of water law and legal wrangling across the West.

RETIRING AND MOVING WATER FROM
THE TOP OF THE WORLD

As of early 2018, seven years after the Aamodt and Abeyta settlements were federally funded by Congress and President Obama, the transfer of retired groundwater pumping rights from Top of the World Farms had passed formal review and approval with the Office of the State Engineer (OSE). It had sat in limbo for years, between the settlement funding in 2010 and the deadline to act on this transfer in September 2017. There was great political pressure to have the farm's water rights transferred from Taos County to Santa Fe County. Taos County tried to protest the transfer, but only silence emanated from the OSE until the approval for transfer was quietly made in late 2017.

These newly acquired rights will soon be drawn from the surface waters from yet another side-channel dam on the Rio Grande. The new intake is to be just north of the Otowi Gage, along the main channel of the Rio Grande on San Ildefonso Pueblo's land (see map 5, chapter 2). Santa Fe County will help administer the new Pojoaque Valley regional water system in a joint powers agreement with the four Indian pueblos to satisfy the requirements in the Aamodt settlement.[5] San Ildefonso may play a more active management role in the future given the location of the intake and pipeworks on its lands. I discuss why this is of particular concern to valley residents in the last section of this chapter.

The water from Top of the World Farms is enough to serve four to five thousand users on an annual basis (see map 7). It is a transfer that legally, but not physically, turns one water type into another: groundwater into surface water. On paper, water rights can be sold and transferred with a pen, but that doesn't make it legally easy or physically real in terms of connected waters. Public meetings as part of the Aamodt settlement reconciliation process took place during the summer of 2011. This "reconciliation" meeting was necessary to match the initial agreement language of the parties (2006) with the language Congress used to fund the settlement in 2010. In early August 2011, my student Andrew and I sat in the air-conditioned chill of the Roundhouse, New Mexico's legislative building. Two other residents from the Pojoaque Valley also sat inside, quietly observing and occasionally raising eyebrows at each other. All of us who were not part of this legal reconciliation process were confused while listening to the lawyers debate, bicker, and redact the reconciliation language.

At the end of the morning session, I asked whether a transfer of Taos water across the Otowi Gage on the Rio Grande might have Rio Grande Compact com-

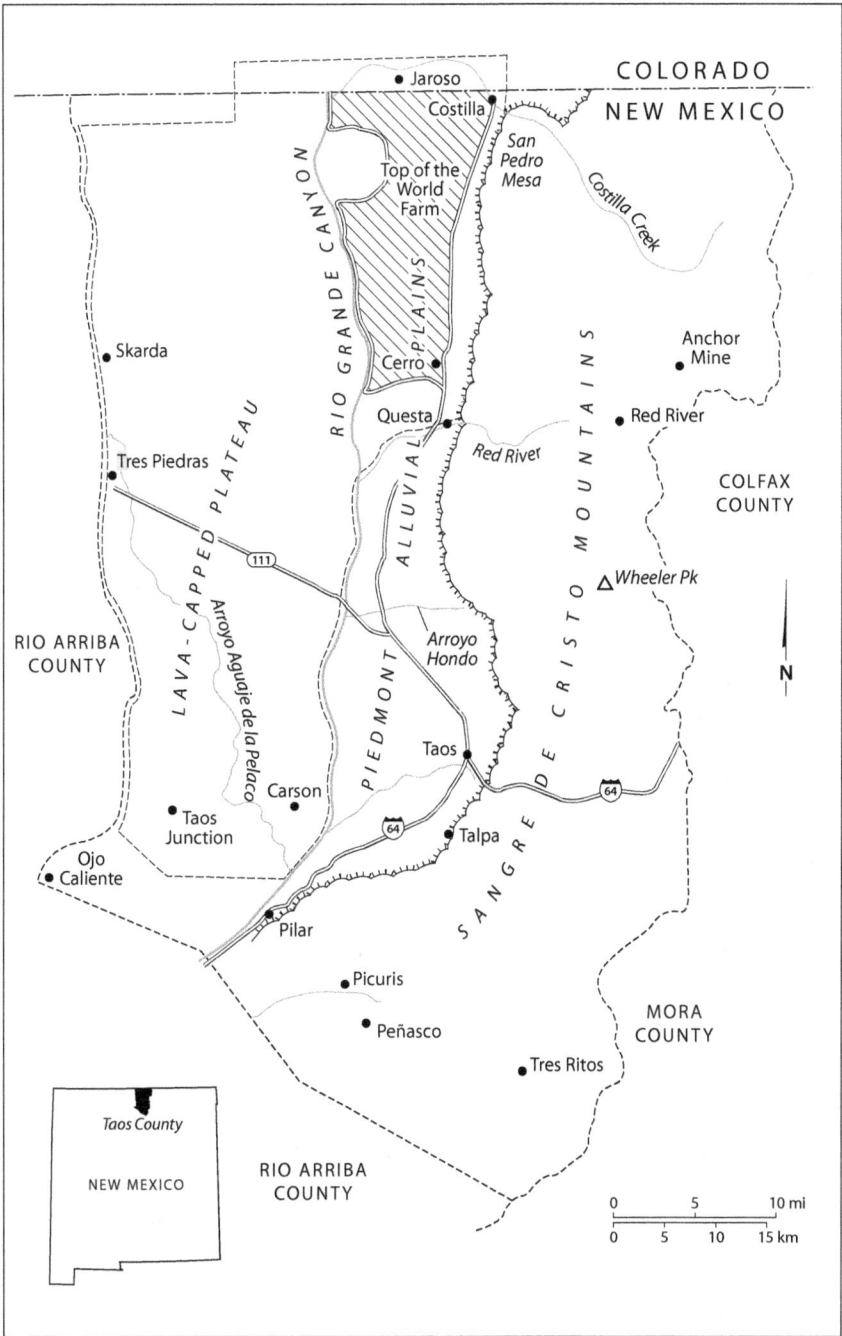

MAP 7. Location map of Top of the World Farms in northern Taos County (in cross-hatching). The purchased and retired groundwater rights from old pivot irrigation are the legal and hydraulic tie between the Taos (Abeyta) and Pojoaque (Aamodt) Valleys. Adapted from the New Mexico Office of the State Engineer and a Taos County map.

plications. Otowi is important as one of the Rio Grande Compact metering points that determine water deliveries to southern New Mexico reservoirs and, ultimately, to Texas. In response, an attorney from one of the four pueblos responded definitively: "That's been settled a long time ago. The OSE can transfer those waters, it's settled, and done. It's no longer in question."[6] While I appreciated the candor, the tone also betrayed how impatient attorneys can be regarding public involvement or participation. That attorney's response also underlined that in these cases, legal and political pressure had to overcome hydrological facts or inconvenient legal aspects between groundwater and surface connections. Interestingly, the proposed withdrawal point for water in this transfer has changed since 2011, and the location of the intake for the new Pojoaque regional water system is now set to be located north of the gauge. This certainly avoids any potential compact implications for water deliveries to Texas.

That 2011 reconciliation meeting also illustrated that lawyers can ignore hydrological complications when they are convenient—or more fairly stated, when a water code gives the state engineer the power to ignore them, and settlement terms demand it. If the water does not actually move, the law can say it does or insist on hydrologic connectivity even where it does not exist. In other words, if the basins are not connected, attorneys and engineers alike pretend they are or devise water equivalency ratios through groundwater-surface water offset calculations. Western water agencies often ask the impossible of actual water to make paper-based water politics possible. It is an openly pragmatic yet covertly political horse trading of water that will never see the surface or the soil of the next basin over. Hydrology is ignored so that law and legal settlement can "work."

That attorney's grumpy response was also an uncomfortable moment. She ended her initial response, too, with the familiar statement about "that's just how it works here." I began to understand how New Mexicans with real questions about settlement can feel sidelined or ignored by negotiators. Water users often feel insulted by experts who hash out these plans with little transparency or knowledge by the wider public. As I mentioned in the previous chapter, a public welfare committee for water issues was organized in Taos County several years ago. On more than one occasion, as the committee met intermittently between 2012 and 2016, some committee members questioned the transfer from Top of the World Farms. Other committee members and members of the city government countered that they should not bring this up. By 2016, even Taos County officials worried that questioning the transfer might endanger the entire hard-wrought yet fragile Abeyta settlement. Legal and political-economic aspects of the Aamodt settlement required the transfer.[7]

The Abeyta settlement, like its Aamodt counterpart to the south, was designed to avoid a senior Pueblo water rights call. These additional waters from Top of the World Farms in Taos County were deemed essential to supply enough surface water for use in the Pojoaque. A statement by a county attorney (J. Utton) working to make the transfer happen expressed a prevailing feeling in the valley. Quoted in the *Santa Fe New Mexican* in July 2015, Utton claimed that the water rights were "as good as gone" and that people should focus on what water remained and protect that water.[8] New Mexicans in both basins and beyond are watching the new regional water system closely to see what will happen with yet another pipe diverting water from the main stem of the Rio Grande.[9] Residents in the valley felt they'd been given little choice in the matter.

As noted in the previous chapters, settlement does not necessarily leave all parties happy. In a way, that is the point, since every party is in theory compromising on a position. I spoke with Doug White, a now-retired hobby farmer, back in 2010. "I feel, honestly, like we settled for a crappy choice in Aamodt,"[10] he said, bitterly. Residents of the Pojoaque and Tesuque Valleys (in the Aamodt settlement area) were rattled by the implicit threat of "settle, or else." The "or else" in this case was that the pueblos might enforce senior water rights over everyone else in the valley, lowering junior well use during drought.

Confusion and tensions about this are understandable when one settled adjudication (Aamodt) depends on water from another (Abeyta), such as the Top of the World Farms transfer. Some critics complained that only select parties got a place in the negotiations and were able to negotiate "under the table" deals. Defenders of water trades such as Top of the World recognize their limitations and inherent illogic but defend the deals as necessary.

As Emert Robinson, an engineer with close ties to the OSE, put it:

> I mean, what choice do we have? Are the two sets of water connected? Probably not. I haven't met anyone credible who thinks that . . . It's just that people had this sale on paper already, and it made its way into the pile of water rights being compiled to enable the Aamodt settlement to happen. I mean . . . politically, it was expedient. The feds funded it. Lawyers say 'Make it so,' and well . . . [chuckling] there it is . . . And we are trying to make it work you know . . . But yeah, the hydrology probably doesn't work.[11]

The Top of the World Farms water rights were the glue connecting both basins. Without an approved transfer from Taos to Santa Fe County, allowing surface withdrawals connected to the Chama/Aamodt pipeline system, it is hard to imagine how the settlements could have worked as currently drawn up. This was not a massive water transfer, yet even this small amount in question was problematic for those cultures of water in Taos that value the local land-water connection. To Taoseños, it was a big deal and viewed as an expensive trade-off in making sure their own settlement happened.

Another complicating factor in these settlements involves the negotiating parties. As noted previously, while the Pueblo were represented by federal legal counsel, the non-Indian defendants were individual parties, meaning there was no possible single tactic for representing, defending, and contesting all of the respective positions held by non-Indian residents. The legal strategies by non-Indians changed to some degree, as seen in the case of the Pojoaque Valley Water Users Association in the Pojoaque and by the Taos Valley Acequia Association for the Taos Valley. Emerging connections between surface and ground waters, even in the settlement language, continued to bedevil both state agents and local water users in these valleys.

The perceived shell game of the water transfer from Taos to Pojoaque understandably sets water claimants and residents on edge. For some, the settlements simply renewed the same old pattern of delocalizing water that was once tied to land in their watershed. In both basins, the signatory parties now stress that it is too late to back out; the deal has been made. When public questions and protests arise, the frustration of attorneys who worked years meticulously crafting the settlement language is understandable. On the other hand, members of the public have been excluded from official roles in the settlement negotiations, and their frustration is also understandable.

The settlements that spin out of adjudications are in some ways no less difficult for relationships between people on the ground. Even federal judges are affected. The judge overseeing the Aamodt settlement recused herself in July 2014, since her husband sat on the Santa Fe City Council, which was a party to the settlement.[12] This was to avoid conflict of interest charges that some antisettlement objectors had lodged against her. Settlement does not simplify. This out-of-court peacemaking process begets additional jurisdictional water fighting and costly engineering, even when local "demand" for more water infrastructure is not there.

Taoseños and Pojoaque Valley residents remain suspicious of yet another water project they fear cannot be sustained either financially or administratively. They view the deal as an unequal exchange, or as one woman from Pojoaque framed it: a "deal to transfer away groundwater rights that we don't pay for to a new plumbed regional system that would be metered for and we'd have to pay for."[13] This is perhaps the most visible and dividing aspect of the new operating plan. Non-Indian residents of the Pojoaque Valley who had been using free groundwater (up to a certain quantity) would have to join a system whereby they have to pay to use surface water, not because their wells have dried up but because their legal standing has changed. Residents are concerned that the state and the county of Santa Fe will escalate costs or do an improper job in governing the regional water system. As the legal and engineering aspects come together, issues of governance, operation, and pricing will keep all parties preoccupied.[14]

One of the most common challenges to both OSE personnel and to local water users trying to understand the procedure is how to calculate water offsets, particularly when shifting a right from groundwater to surface water. Both Pojoaque

Valley residents and Taoseños are anxious about limited groundwater rights, increased pumping from new mitigation wells, and the idea that aquifer storage recovery projects will meet all of the needs of their valleys. Settlements should be viewed, like adjudication, as a process even after the fact, not a one-time clean-cut deal that ends all disputes. They take ongoing work and continued discussion and collaboration to work. Built in to these conjoined agreements are some important assumptions about water and water rights.

SETTLEMENT AND ITS DISCONTENTS

The Abeyta settlement includes set-aside funds for hypothetical hydrological projects that are designed to benefit multiple parties: "x many acre feet will be acquired from the Rio Grande main stem, y many acre feet will come from San Juan Chama water, and z acre feet can be acquired with the new federal funds to buy future water rights for mutual domestic associations." These are all framed in the future or conditional tense, as a way for the parties to feel they will all benefit in some way if they stay in the agreement. Since adjudication offers no financial benefits or "new water" in either valley, settlement seems preferable, even if the terms of agreements may be difficult to fulfill. That promise of future water infrastructure to balance the demands of groundwater and surface water trade-offs is powerful.

As a member of the Taos Valley Acequia Association said of the Abeyta settlement during a public meeting in Taos in 2015, "It's still a damn good deal, though I was the hothead in the room who walked out on numerous occasions from those meetings." Yet in that same meeting, as discussions were coming to a close and people were congratulating the settling parties in Taos, one senator asked for "an update of sorts, in a year, to see how the acquisition of surface waters on the Rio Grande main stem are going, because . . . it kind of sounds like wishful thinking, and I think we deserve to hear whether this arrangement will work long-term." You could hear a pin drop in that awkward silence, right before Chairman Wirth promptly ended the public meeting session.

To be clear, moving away from groundwater is desirable and logical in places where groundwater tables are dropping rapidly. For decades, the cities of Santa Fe and Albuquerque had been pumping groundwater unsustainably, and moving to surface waters provided by the San Juan-Chama Project made sense. Depending on more renewable surface supply is more sustainable. Both Taoseños and residents of the Pojoaque Valley are well aware of lowered aquifer and groundwater levels.[15] Yet concerns remain about this laudable groundwater-to-surface supply shift and whether the main stem Rio Grande will actually have enough water for these bartered water quantities. All of the parties, in both settlements, are betting on the same horse: the Rio Grande and its San Juan River-augmented hybrid waters.

The Abeyta settlement provided funds for the town of Taos, domestic mutual associations, and Taos Pueblo to buy surface water rights as they become available. The terms also specified that the pueblo and the town would get first rights to purchase retired agricultural water rights when those become available. Getting residents of Taos County to buy into these technical, hydrological, and political arrangements has been a struggle for the state engineer, the federal Bureau of Reclamation officials, and their nonadjoining neighbor to the south, the county of Santa Fe. Abeyta also makes some serious assumptions about the future availability of surface rights to meet the obligations of the settlement expectations. A long-term monitoring and transparent plan to verify water rights sales and transfers will likely be necessary for these settlement terms to be successful.[16]

The groundwater and surface water games of the state continue to befuddle many in both basins, for good reason. The OSE, attorneys, and hydrologists treat that water differently. The groundwater is retired, then a portion of that retired groundwater amount can be taken out downstream off the Rio Grande. Water is not magically levitated from Taos groundwater and lowered into the Pojoaque Basin as surface water. "If that seems weird, Eric," a hydrological consultant told me, "well, that's just the way it's done out here. We have different ratios for calculating how much ground- and surface water can be offset in these transfers. Without them, we can't really move water at all, and no one's going to be satisfied with that. These agreements fall apart without that mechanism. It seems strange but that's how it works on paper, at least."[17]

The parties involved in the settlement realize that not using water in one place and using water where it is actually extracted is a sleight of hand, even if some refuse to admit it. Today's frustrated Pojoaque Valley residents are learning once again how settlement may not actually "settle" the troubling cultural relations in their valley. As Tina, a resident of the Pojoaque with whom I spoke in Taos a few years ago, summarized it, "We feel, honestly, that the pueblos are agreeing to be 'used' by the county and city so that the pueblos can get as much water to then turn around, and then lease it back to the county and the city. But no one can say it to our face; no one is being honest about why water leasing is part of all this settlement talk."[18] In a system where trust and visibility have always been a part of the fabric and moral economy of the valley, one wonders if that trust will endure.

For the Pueblo parties involved, it is hard to underestimate the importance of the settlement terms and what they are getting. A senior member of Tesuque Pueblo, Tony Herrera, explained to me as such:

We signed that accord in good faith, and it's important to all of the Pueblos to have assurances and guarantees that we have first rights to water in the valley. We aren't playing games with this; we want to get along with all of our neighbors but it's difficult. The county has laid out roads that go through our ancestral lands, and they want to lay the pipe down there along those roads. Then non-Indians turn around and complain when we want to claim those areas, or the county gets mad when we question the pipe easements. There's nothing simple here about all this, and we probably need some federal intervention to calm down the property disputes. A lot of non-Indians bought land along those roads that are Pueblo property too, and they are worried their property values are going down because they have fewer groundwater rights? I mean, this was our land, and they came and took it. Now that we want some of it back, and some guarantees on water, everyone is upset. It's ridiculous. We should be realistic that everyone has given something up over time, you know? And just try to tell me we haven't given up the most. Most of this valley is not in our hands anymore. And everyone else is screaming injustice?! Please.[19]

From Tony's perspective—and that of tribal councils and their attorneys—these water settlements were perhaps one of the last remaining leverage points to get some resource justice; water was the last legal card to play in a region where the Pueblo lost most everything else.

Water rights are not absolute rights to water, technically, and the terms of Aamodt and Abeyta remind everyone that this is the case. But try telling that to a property owner who thinks the state, the Pueblo, and the federal attorneys "took" part of their groundwater rights away. Owners of wells installed after January 13, 1983, face restrictions on outdoor water use as part of the settlement. Owners of these post-1982 wells feel like their property values are thus depressed by this lowering of absolute well water use.[20] The OSE typically approves domestic wells, and in the past the amount of water rights awarded was predictable, seemingly stable. The state engineer's permit power is now limited by this multiparty (Aamodt) settlement and caps the total effective water rights that households in the Pojoaque can use.

No one wants further litigation, but the net results of these agreements have clearly been contentious. Bob and Jacob were still angry when I caught up with them at a meeting in 2016. "Sure, we settled," said Bob, "and not in a good way, mind you!" Much of his ire fell upon the experts then negotiating "our water future," as he put it. Jacob echoed his neighbor's feelings. "These lawyers were just, you know, determining how much water people get without any kind of state engineer input, or talking about how these water rights were getting curtailed because of one miserable judge back in the 1980s . . . how is that justice?"[21] As the consequences of these agreements materialize on the ground for the parties, the water compromises built into the language are becoming all too real.

LOCAL WATER PROTECTIONS AGAINST MOVING WATER

The water transfers remain a real local concern that the OSE will have to navigate long into the future. Transfers immediately affect the local system in question if the terms of the transfer are honored (locally). In the long term, it is the OSE that will have to balance the water spreadsheets of surface and groundwater equivalencies across these basins, since there is no statute on limiting interbasin water transfers or equivalency formulas. These political and legal wrangling mechanisms for water transfers to settle lawsuits concern rural irrigators. Oddly, if adjudication was what many New Mexicans feared might move water, these two settlements did the actual work of moving water from one basin to another. Amid these new concerns, however, acequias and other ditch organizations are not powerless.

Acequias have tried to carve out spaces of exceptionalism in the New Mexico water code, such as the newly legislated ability to deny a water transfer or sale away from a ditch. In many cases this has been remarkably successful. They have in effect claimed and executed a kind of *settler indigeneity*, one based less on indigenous identity and more on time in residence and rural patronage political clout, to substantiate their defense of local water.[22] Because of the lack of recognition for communal claims to water by the state, trying to carve out water policy exceptions is one of the few tools available to the older ditch organizations in New Mexico.

Pursuing this strategy of "acequia exceptionalism" to evade state water policies can feed a perception of cultural favoritism toward the acequias, which are not obliged to follow statutes that affect other mutual ditch and water-user associations. Or as one owner of a dairy farm located south of Albuquerque put it, "They get all these little clauses that give them the ability to ignore changes in law that the rest of us have to abide by; it starts to build some resentment towards them. I mean, we have to follow the law, but they don't? How is that fair in the long run?"[23]

Another important concern is internal to the ditches themselves: their relative capability and liability. This comes to a head when acequia commissioners deny individual (parciante) water transfer proposals that would move water rights out of the acequia's jurisdiction. Commissioners often feel forced to participate in this system that begins to "act like a kangaroo court, in my view," as John L. from Taos shared with me back in 2010 at one such water transfer hearing. Being able to say no to your neighbor and fellow parciante also carries a burden of proof. Commissioners have to be quite careful about giving fair due process to those who wish to sell and transfer water rights. Commissioners have to be able to defend any eventual decision based on the weight of testimonial objections from other water users that drawing water away from the ditch would harm remaining parciantes.

The seniority of individual water rights along acequias contributes to their attractiveness; these rights are targeted by wealthy, willing buyers. The sale of senior water rights can affect all water users on a ditch in terms of water availability and flow. However, the denial of a transfer could also harm the ditch by harming interpersonal relationships. After all, it is difficult to prove that moving water out of the ditch would impair others beyond a reasonable doubt. But so far, the statute has proven valuable in keeping water in ditches and in keeping water as local as possible. These new exceptional powers for acequias, then, for avoiding water loss and denying water transfers come with costs that increase their exposure and liability.

John from Taos shared these doubts and whether they were simply going to face more liability and litigation: "It really adds to our level of . . . bureaucracy and makes us all nervous about pretending to be our jurors and judges. I mean . . . [he sighs] we have to do right by the other folks on the ditch, too, who fight tooth and nail to keep water on the ditch. It just puts us in a bind we never thought we would have the power to be in."[24] The problems with this new system are easy to imagine and difficult to reconcile. Ditch users become pitted against one another. Individual rights are not always the best solution in a system that has traditionally balanced the whole needs of a community. However, the argument that acequias be given the water rights as entities, rather than rights to individuals, has fallen on deaf ears in the courts and with the OSE. Modern water law denies collective or communal water rights, just as courts refused to recognize common property land grants held by Hispano villages between 1848 and the 1920s. US jurisprudence wants bodies and individuals, not a commons, or a collective.

The OSE has, over the last thirty years, become more amenable to awarding a single ditch-wide date for acequias. This happened in the Pojoaque, just as it has in the Jemez Valley of New Mexico. Ditch-wide shared dates create fewer fights on any single ditch, even if different ditch systems have different dates. This is also more realistic for water management should prior appropriation be invoked, as then OSE or a special "water master" would only have to sort the different dates between ditch systems rather than every single water right if different dates were held by different individuals. Culturally and administratively, awarding a single priority date for an entire ditch made sense to both local users and eventually to the OSE. As long-time journalist V. B. Price put it, "When it comes to water, there can be no winners and losers. We all need water. The political realities of New Mexico don't permit priority calls, even if such calls are legal."[25]

What does this have to do with the settlements discussed here? To parciantes and mayordomos, the danger of these water agreements is that they are connecting waters in ways that are nonlocal, yet again, to satisfy the demands of too many parties spread across hundreds of miles. To rural irrigators, the new language of water market "efficiency" going to the highest economic use is simply rhetoric and suspect. For those who adjudicate or will have to engage in further settlements, the new protectionist features that acequias possess are potentially of concern if water cannot be easily renegotiated or moved across or between ditches. The new acequia legal protections are cementing highly local terms for water use and access, further removing them from the kind of flexible connectivity that might be necessary for water settlements. New bylaws have kept water more local but are increasingly tested by individual water rights owners almost every year on small ditches across the state. And the bylaws do little to reconcile the new challenges occurring between surface and groundwater users.

INFRASTRUCTURE TO ADJUDICATION TO
SETTLEMENT TO INFRASTRUCTURE

There is still a small but fierce protest by parties who feel cheated by the Pojoaque settlement. The settlement's terms particularly affect non-Indians with groundwater wells in the valley, as discussed in chapter 2, especially if the wells are post-1982 groundwater domestic wells. Not everyone wants to be connected to a new water system that would be run by Santa Fe County in a joint powers agreement with the four Indian pueblos. The primary resentments stem from individuals feeling that they are being forced into a new system in which they have little input regarding governance and that they will have to pay for water they once got for free from wells. In the end, the loss of water microsovereignty over a well and paying someone else for new surface water that was once free well water are both problems.

The new regional water system is being funded by the US federal government and the state of New Mexico. The costs by 2024 are likely to soar beyond the $261 million estimate being shared recently at public meetings on the issue.[26] This will secure the surface water supply to the non-Pueblo residents of the valley who choose to connect to this new system. This planned regional water system for the Pojoaque Valley is slated to have an intake on the main stem of the Rio Grande, just north of the current Buckman Diversion Project that feeds the city of Santa Fe (see map 5 in chapter 2 for reference). Again, this makes good hydrological sense, reducing well water use and providing a more reliable surface water supply.

However, there are also real hydrologic questions as to the ability of the single stream of mixed waters in the Rio Grande to supply the entire area with drinking and municipal water. Everyone downstream and upstream takes a share of the mixed Rio Grande; some of the water is the Rio Grande's real river flow, but much of it is also San Juan-Chama Project water. As separated rights to these conjoined waters accumulate and real water withdrawal infrastructure is built on the ground, it is hard to separate the waters and see how everyone will get their fair share.

Plans for the new Pojoaque regional water system, part of the Aamodt settlement, are being challenged. Around seven hundred public objections have been raised about the proposed water system and who will manage and govern the new water deliveries. Initially, the associated water plant was to be managed by San Ildefonso Pueblo. The overseeing governing body for the water authority was to be composed of five members, four from the respective four pueblos and one at-large or county representative.

The Pueblo people make up only 10 percent of the population in the basin. Non-Indians in the valley protested that plan, arguing it gave too much governance authority to the Pueblo, who are in the minority in the valley. As of 2016, the plan was changed to a body of nine members, allowing for more non-Indian representation. Santa Fe County will administratively govern the project, at least at first, despite much skepticism about this new water system in the valley.

Just as engineers have designed transbasin diversions across the western United States in the twentieth century, legal language in the settlements to satisfy both the Aamodt and Abeyta water agreements begets new engineering. These agreements are perhaps better framed as engineered negotiations, with law and engineering recruited again into the postadjudication process.[27] Law and engineering coproduce each other and are not separate entities in adjudication (as I discuss later in chapter 6). In the twentieth century, watersheds were simply a set of bathtubs to be connected if water was in the wrong place for human use. Engineers found ways to move water across those basins and did so with remarkable effectiveness, although the consequences were often unclear.

As New Mexico engages in further settlements, officials, irrigators, lawmakers, and adjudicators should pay careful attention to the successes and failures of other water settlements in the western United States. Even settlements tailored to particular situations include lessons that can be transferred to other settings. However, trust and social capital is crucial and cannot be abstracted or treated as just a template. Trust has to be in place or created through the long process of meetings and negotiations. In a way, settlement is a highjacking of the legal phase of adjudication itself, minimizing the inter se dispute phase. Settlements are increasingly taking away the decision-making role of the state courts. However, state and federal funding and the OSE are still just as vital in these agreements. Some court oversight, as seen in the mandated benchmarks in the Abeyta case, remains intact. Agreements, with all their legal challenges and new infrastructure problems, serve to reconnect what adjudication socially severed. There are cautionary lessons here for other western states that might treat settlement as a one-size-fits-all template for solving adjudication problems.

Dams and water projects drove the adjudication process in the twentieth century. Legal agreements and settlements are driving new diversion and pipeline water infrastructure in this century. The old twentieth-century "iron triangle" of congressional representatives, agricultural lobbies, and agencies is far from dead in this century.[28] These less obvious side-channel diversions will also have hydrological consequences in times of scarcity and produce river channel changes over time.[29] New projects are difficult to carry out, considering how the cultures of water in New Mexico have long disagreed about the values, purposes, and measures of water. Adjudications did not invent new water conflicts; they simply revealed long-standing water antagonisms between the various water cultures by trying to extract a single "core" of water as a resource from these multiple cultural layers.[30]

The new connectivity of western waters, driven by legal agreements through settlement, also poses new challenges to both local and expert forms of water measurement across the state. Federal and state agencies will seek to provide water from any available source, while local irrigators and larger irrigation districts will strive to cement legal protections to keep waters local. Where adjudication failed because it fragmented cultural understandings of water, local parties took back some power through settlement. Paradoxically, as old notions of social capital and water sharing were built into the agreements, the same old ideas about infrastructure and dependency on federal and state funding returned. The western states and the federal government continue to use the "Indian blanket" treatment in water settlements. This is clearly visible in these latest arrangements that drive further infrastructure and water delivery. Groundwater, surface water, water sharing, and water sovereignty are now all conjunctive, bundled into the final agreements, like it or not. Whether these latest arrangements actually end up delivering new infrastructure and wet water to the pueblos will be worth monitoring. As McCool gloomily assessed Indian water settlements more than fifteen years ago, "Indian tribes won many court battles, but Anglos nearly always ended up with the water."[31]

These basins and cases illustrate how complicated the situation is in terms of infrastructure, adjudications, and connecting water sovereigns. New Mexico's share of the Colorado River Basin, across the Chama and Rio Grande Basins, required connective infrastructure. The San Juan-Chama Project forced the state of New Mexico to sort out "project" water versus "native" instream Rio Grande water. This meant adjudicating multiple New Mexican water basins. The results cascaded into the Aamodt and Abeyta adjudications and into eventual settlements that required moving project water and groundwater to settle these claims. Infrastructure led to new infrastructure, in short.

Importantly, the two settlements *demanded* that hydrological connections exist even when groundwater and surface water users may not be connected. As part of this conjoined water process, experts played a key role in driving a new calculus to thinking about water in abstract measures. The new universal and abstract measures of water were largely foreign to New Mexico and the West just a century ago. The development and deployment of these new metrics were vital to the development of expertise as adjudication was happening in the state. Part 2 examines the consequences of the new "expert" approach and regime of water management. The deployment of new measures, new water experts, and new water groups formed in New Mexico as a result.

PART TWO

The Production of Water Expertise: The Adjudication-Industrial Complex and Its Consequences

5

Changing Measures

How Expert Metrics Change Water

No one of course knew, back then, right, that the station there on the Rio Grande [Embudo Station] would be so important. It was the first time we had seen these white guys with hats on trying to measure water, sticking in rods, and strange devices . . . I mean that's what my grandfather and father told us about, and yet we had no idea that that, right there, was the beginning of this obsession to change our measurements, to abandon our system of measuring with surcos and time slots for irrigating . . . to go to these strangely alien system of acre-feet and cubic feet and miners' inches and all that weird stuff. We just didn't think that way . . . we do now, more commonly you know, but back then [waves dismissively, frowning] . . . it was just another imposition, even if we had no idea at first what they were doing. They didn't even know what they were doing at first. . . . apparently they lost a lot of instruments trying to measure the flow. That's how it is—she [the Rio Grande] still calls the shots, and she's still in charge.

—JUAN ESTEVAN ARELLANO[1]

The state's self-imposed demand for water adjudications created much that was new under New Mexico's sun: new and larger infrastructure, renewed conflicts over water, and most potently, new ways of thinking about water as a form of property. Paradoxically, little expertise existed to carry out the mandated adjudications at first. Thus, a whole new culture of water expertise had to be created. Since the late nineteenth century, new agents and experts in New Mexico have been tied to particular places where that expertise was developed. Such places, like Embudo Station on the Rio Grande, were the site of creating new metrics, measures, and benchmarks for particular practices such as flowmetering. These spatial and hydraulic expert places demonstrate how much of the work of experts converges on specific sites of technology. Early water measurement sites—like the Otowi Gage or Embudo Station on the Rio Grande—were key places for developing objective, quantitative "water expertise" in New Mexico and throughout the

89

western United States.[2] If the state was to redefine water as something other than communal or a shared resource, experts were vital in developing new metrics for this new culture of expertise.

State officials further coordinate with a phalanx of federal hydrologists, modelers, and dam operators who account for and coordinate the macrowater in the state's flows. The first step in developing this cadre of water experts depended on transforming the measures and metrics for water itself. These metrics were necessary for the state—and the federal government more broadly in the American West—to produce a more scientific or modern notion of water. These measures, experts, and metric technologies, in turn, were instrumental in changing notions of water sovereignty and asserting territorial and state engineer control over New Mexico's waters. However, producing and inserting water experts into the process did not necessarily simplify the lives of water users.

To achieve adjudication, the Office of the State Engineer (OSE) required new measures that could make water use across the state commensurate. Local waters had to be translated into technical water and access into a private-use right that could be individuated. The OSE did not go it alone in this work. Other federal agencies, state institutions, and disciplinary expert practices were instrumental to the rise of "modern water." The new metrics necessary for adjudication expanded in use beyond that initial purpose in the twenty-first century. The OSE implemented new metering and monitoring, including the Active Water Resource Management rules discussed later, when the 2002 drought provided an entrée into state management of water even before adjudication was complete. Here, I explore how metrics and measures that shape the management of water in New Mexico were negotiated and ultimately accommodated. I end the chapter with a contemporary set of examples that demonstrate why measures, meters, and new metrics are still contested by local water sovereigns and how they allow for new water expert insertion points.

EXPERT METRICS MEET CUSTOMARY MEASURES

In 1889 the United States Geological Survey (USGS) installed its first-ever streamflow gauge.[3] The gauge, referred to in the chapter's lead quote, measured Rio Grande flow near the village of Embudo in northern New Mexico. It's still there today. Under the leadership of John Wesley Powell, and under Clarence Dutton's direct choices for location, a new generation of hydrographers trained at Embudo's "camp of instruction." Under the civil engineer Frederick Newell, the camp developed methods and instrumentation for standardizing streamflow measurements. Newell's corps of hydrographers went on to work across the western states, establishing water codes and solidifying their new sciences, later known as *hydrology* and *hydraulic engineering*.[4] The invention of H2O as a modern substance was dependent on specific places and people.[5] Although expertise was presumed at the instant of creating water law and adjudication's demands, it was still a work in progress. It took decades for hydrology and hydrography to develop a body of knowledge to catch up to the legal demands of the state for adjudication practices.

The newer versions of these technologies, and their related, more humble brethren (such as simple flowmeters), are still fundamental to water experts and expertise. Without them, there would be no data to make state compacts work. Without the gauges, there is no empirical basis to track the allocation of stream-flows. Without the data, no modeling can occur to create the vast flowcharts used by dam operators. Engineering and technology depend on these data and devices to make the legal infrastructure of western water work.

The importance of technology, instruments, and their places is such that when they reach milestone ages, they get their own commemorative status as monuments. Embudo Station, for example, celebrated its 125th anniversary in 2014. The state engineer (Verhines); the acting director of the USGS; and state, federal, and local officials were all in attendance. The ceremony was organized "to recognize the device that set the foundation for modern water management."[6] Tourists, to this day, can stop and admire the country's first water gauge, marked by a rather grim, gray monument and plaque at a roadside pull-off on the east bank of the Rio Grande.

As science, expertise, law, and hydrography became intertwined, the western states were still struggling to reach water-sharing accords among themselves. Such accords were largely accomplished through complicated and drawn out river compacts, all of which depend on water measurements and monitoring. New Mexico shares the main stem of the Rio Grande with Colorado upstream as the headwaters and Texas and Mexico downstream. An unassuming yet crucial technological object sits in the Rio Grande just south of Otowi Bridge, on the road between Santa Fe and Los Alamos. The Otowi Gage measures and registers flow, which then determines how much water must be delivered, per the Rio Grande Compact, to Elephant Butte Dam and Caballo Reservoir for Texan use downstream. Such state macroprojects often came about through interstate threats of litigation and had to be dealt with before completing inner-state water accounting via adjudication. The river compacts are to interstate water relationships what adjudications are to each individual state in the western United States.

Embudo Station and Otowi remain important. Embudo provided the training grounds for experts, and Otowi exemplifies how a gauging station determines "macrowater" sharing with Texas. Through these places, measures, and their new instruments, the new culture of state and federal experts could begin to understand the large units of water flowing in and between the states. New Mexico agreed in the 1938 Rio Grande Compact to deliver a fixed percentage of the flow going by Otowi Gage into Elephant Butte Dam and irrigation districts down south on the Rio Grande. This had repercussions, and some ruefully joke that Texas thinks its boundary begins at Elephant Butte. In dry years, New Mexico often struggles to meet the defined terms of water delivery included in the compact (see chapter 8 for more on this).

On the ground, the rise of expert measures coincided with the rise of prior appropriation water law predicated on an individuated "microwater" legal structure: individual water rights. Water rights needed a quantitative metric to then be individuated on the detailed OSE hydrographic maps. The hydrographers made their measures in cubic feet and meters per second, and individual water rights were doled out in acre-feet of water per year, abbreviated as AFY, a relic of nineteenth-century western homesteading and agrarian settlement policies. An acre-foot is equivalent to 325,851 gallons, roughly enough water for four households for a year.

However, many water users, from acequias to existing cities, found the measure of their water rights as confusing as the state's redefinition of water. The units were incommensurable with local understandings of water and often conflicted with already functioning ways of measuring water. When well-accepted and understood local notions of water management and allocation are questioned and changed, things can fall apart.

WATER PERFORMS A DUTY

New Mexico had already experienced an earlier clash of worldviews over measurements centuries before. One can imagine the Pueblo Indians watching as Spanish surveyors moved about fields with the vara (measuring stick), surveying and mapping arable land for colonists (see figure 5). Those medieval Spanish units were foreign to the Pueblo at the time, and their imprecision and variability also led to much litigation throughout the larger Spanish Empire.[7] Measurements matter and new, imposed measures often transform the existing cultural landscape and human relationships.

Similarly, mayordomos in the twentieth century warily watched as hydrologists and later technicians arrived in their rivers and fields, measuring water flows and mapping out water rights in new foreign units. One of the technicians' jobs was to determine the "duty of water" as part of the technical step in adjudication. The duty of water is a climate region–specific average, in acre-feet per year units, for specified crops that are growing at the time of the surveys. Grain crops take less water per acre, alfalfa requires more, and tree crops need even more water per acre. For local experts, this was problematic. Streams changed from year to year, depending on snowpack and precipitation. So did crops, and ditch bosses and irrigators were more concerned that enough water would get to their fields than about "duties" set for a single crop at a single point in time. Mayordomos allocate

FIGURE 5. Photo of the statue of Pedro de Peralta and his colonial surveyor's vara measuring stick. The statue is located outside the US Postal Service office in Santa Fe, New Mexico. Photo by the author.

irrigation water based on the size of the plot to be irrigated. In dry spells, mayordomos set limits on when and for how long irrigators can use the communal water (often in two-hour increments).

New Mexicans now generally think of their water rights in acre-feet, the standard metric set by the 1907 water code. Coincidentally, the flow through most ditch headgates typically amounts to about one cubic foot per second (cfs), or two acre-feet. This would have been clear to newly formed irrigation companies or federal irrigation districts even back then. However, for most of the last century, irrigators along acequias thought in terms of *surcos*. Old-timers today still do. Old and new measures—like definitions of water—have overlapped and sometimes conflicted throughout the adjudication process.

Surcos refer to swells of water moving through a ditch. A surco is described in hand gestures, indicating how deep, wide, and strong water flow has to be in a particular ditch to be considered a swell.[8] Juan Estevan Arellano explained how the surco could be broken down into finer measures for calculating a parciante's water "share" along the ditch: half a surco, a full share, two surcos, and so on.

A surco is not a set volume. It varies by ditch, and irrigators only a few miles apart could have different understandings of the measure tied to their own ditches.[9] In its etymology, the term *surco* takes its root meaning from the Latin word for *furrow*, showing its pragmatic links to land and flood irrigation.

Typically, mayordomos allotted surcos less in volume and more in time rotations, based on how much water was needed to fully cover an individual field and the irrigator's needs and priorities. The advantage of this system is that it inherently created enough hydraulic force to irrigate entire fields. Today's system of individual rights in acre-feet does not explicitly do so. The acre-foot system is a universal measure, blind to particular field needs or uneven topography.

So the use of water rights not only varies in quantity and nature but also differs in cultural-scientific norms of measurement. This is important. While the stakes of this difference may not seem crucial, there is an important distinction between how much water flows through a total canal and how much water irrigators are allocated to use from that ditch. The mayordomo, or parciante, does not really care how many acre-feet are moving through the canal. What they care about is that water reaches all parts of their planted field, or their garden, or their livestock tank. Mayordomos allocated water based on its availability, weekly and seasonally, rather than by some universal measure. Volumetric water arrangements for the duty of water for a single field did not include such considerations. Furthermore, even today's theoretically calculated duties might not actually match crop consumption rates.[10] Surcos were expressly pragmatic, not fixed or abstract.

These abstractions and measures of water matter to this day. Mario, one of the former defendants in Aamodt on the Pojoaque, noted how law used expertise and these new metrics for a new kind of artificial precision from water:

> That judge [Mechem], he was so stubborn in the [Aamodt] adjudication, he just treated the entire area, all these rivers, like he had a cookie-cutter in his hand, and he demanded numbers from every cookie, even that water was moving between our lands, between whole villages. It couldn't be counted. He was like trying to count water with a net, scooping faster and faster thinking that would work. But that water escapes, it moves, it hides, it goes from field to field. The whole system of you and me and water [individual water rights counted], it just doesn't make sense. It just moves across our fields, we drink, we recycle it [snickers], and we have to continue to share it downstream.[11]

The duty of water was about the fixity and universal principles embedded in water law, as the example shows. Assigning a duty of water (much like a water right) implies a static quality to water flow that does not exist. In most years, streams and canals do not carry the average streamflow. Flows wane and surge and rarely coincide with a calculated average. For example, Joseph may have three acre-feet times ten acres, or 30 AFY (acre-feet per year), of water duty. However, in a low-flow year, he might not get enough water. Local ditch irrigators long knew this even if they did not talk in terms of hydraulic head.

The hydrographic maps drawn by the OSE fixed a static duty (quantity) of water rights attached to a static landscape of ownership (acreage). This is reasonable in that it would be difficult to imagine a similarly predictable, logical way of mapping land and water appurtenance, especially as earlier technology had no way to convey dynamism in water fluctuation. Old adjudication hydrographic surveys created a water-use snapshot of the state to acquire basic figures and a rough understanding of how much water was used in any given valley. Yet no one using water at the time was using a static or "average" notion of water flow along any of the rivers or canals in volumetric terms.

Nevertheless, there may be an upside to how the OSE allocates water duty to land parcels. The amount of water duty assigned is usually plenty on an annual basis, even if the pulse or flow may have to be aggressive to reach the ends of a hypothetical field. Few irrigators used the full duty in the past. Additionally, agricultural acreage in New Mexico has declined. Thus, water duties on the upper watersheds may not be completely out of sync with available water. Furthermore, water makes its way back into the larger stream system as groundwater percolation and *tail water* straight to streams. The amount of water accorded per acre, as a constant, may actually help during scarcity. Since New Mexico's adjudication process, unlike other states, does account for consumptive use rather than just total water right quantities, it produces a more accurate accounting of water use. This makes New Mexico's approach distinct and worthy of attention.[12]

Calculating this consumptive use of water as part of adjudication is vital for the state. The consumptive part of water use is the only portion of a water right that can be legally sold and transferred elsewhere. This metric is critical if the OSE is to oversee and manage active transfers of water rights across the state in the future. If a water rights owner has 100 AFY, for example, and 50 percent of that is direct consumptive use, he or she can only sell away the rights to use 50 AFY. The rest is to stay in the stream system so that others downstream are not impaired by such a transaction. There are active water markets across most western states, and most agencies have developed calculations for transfer (*conveyance*) losses, seepage rates, and so on. Other water rights holders on a system can contest these consumptive or conveyance loss numbers when a sale and transfer is proposed.[13] These sales, leases, and transfers haunt local ditches. The reality is that these will become more common, especially in times of water scarcity. Cities, drought, and suburban development can all factor in pushing the price of water upward, creating more temptation to sell water rights away from ditches and farmland.

Because of the limits of static maps and long-delayed administration of water in New Mexico, it took a severe drought to challenge the technologies and institutions the state had created during the twentieth century. Mayordomos and parciantes had long coped and struggled with water shortages. But the new meters and metrics allow the OSE in New Mexico to insert itself into water-user arrangements on the ground. Among the first measures implemented by the OSE to cope with drought were a set of rules on *active water resource management*, as it has come to be termed in New Mexico. This affected a variety of basins once put into place, and I discuss the implications of these measures next. Expert measures continue to exert their influence and allow for state insertion into water governance to this day.

WHEN MEASURES CONTINUE TO MATTER

The severe drought of 2002 stressed ditch institutions in basins around the state, with disputes erupting across stream systems as irrigators faced scarcity. That same year, senior water rights owners tested the OSE's ability and willingness to enforce prior appropriation law. On some ditches, only the senior water rights holders received water, much to the consternation and resentment of neighbors with junior rights to that same water. Not since the drought of the 1950s had the state been so severely tested for allocating water fairly.

At the height of the 2002 drought, State Engineer John D'Antonio responded to complaints regarding water disputes on ditches. He directed the OSE to propose modifications to the New Mexico water code so that the OSE could intervene and not only count water but also manage it. The program was called Active Water Resource Management (AWRM; "a worm," as it is typically pronounced in New Mexico), and the state legislature approved it quickly. It gave the state engineer broad new powers. A number of basins are now under this new program.

To some irrigators, the solution has been more upsetting than the original drought and its challenges. Much of the blowback has to do with AWRM's main components: new meters for monitoring and "special masters" with broad power to control flows and settle water disputes. New water flumes and water meters were installed throughout the basins with disputes. Special masters and river masters were, as assigned by the OSE, also given powers to administer water using prior appropriation rules, even in *basins that were not adjudicated*. From the OSE's perspective, these special masters would serve as a neutral third party to help settle disputes or long-term water challenges. Typically, these individuals have some background in water resources, although their level of expertise is highly variable. They are also not necessarily from the basins they serve.

The AWRM program highlights two important aspects: first, the obvious role of new experts in the new metered vision of water, and second, AWRM was necessary because adjudication was taking too long, and the state engineer felt intervention

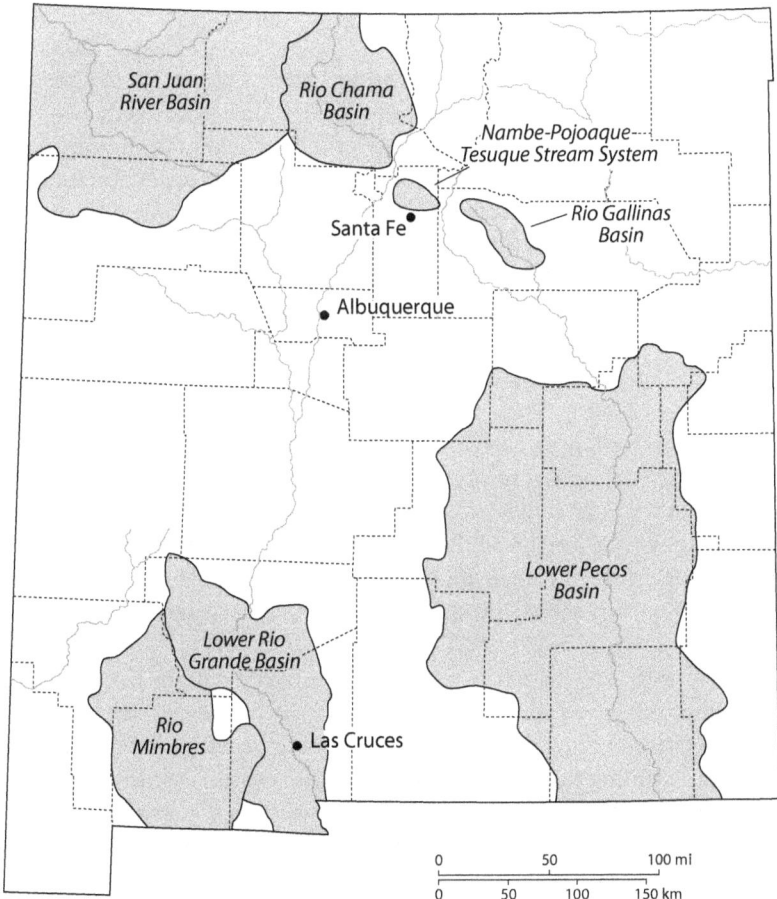

MAP 8. General location map for the Active Water Resource Management (AWRM) basins currently administered by the Office of the State Engineer. AWRM measures were instituted after the 2002 drought, with new legislation. The rules allow for special masters to be designated for each stream under drought duress. AWRM can happen before, during, or after adjudication. Adapted from the New Mexico Office of the State Engineer.

was necessary during times of water scarcity. Despite challenges to the AWRM rules by water users, a state court in 2012 supported the new measures and affirmed the state engineer's broad administrative power to allocate water in basins that have not yet been adjudicated. Here, I examine two basins and their experiences under AWRM: the Rio Gallinas Basin, which includes the small city of Las Vegas and its acequias, and the Rio Mimbres in Southwest New Mexico (see map 8).

"A NEW WATER SHERIFF IN TOWN"

The Gallinas feeds into the larger Pecos River drainage and is still under active court adjudication for water rights (see map 8). This area of northern New Mexico has struggled to equitably distribute water during the last fifteen years of water shortages, especially between the city of Las Vegas and the regional acequias.

"We didn't ask anything from the state engineer," Antonio Vorán said bitterly. He was from the Gallinas watershed and a member of an acequia. "I mean . . . who would want the engineer involved in this . . . what's he going to do, make it rain!?" Some Rio Gallinas water users had called the OSE in 2002 requesting help, but as Antonio said, they were only looking for temporary assistance, an informal and seasonal intervention, much like a mayordomo's informal and seasonal duty on a ditch. As he put it: "Sure, there were fights between ditches at the time, and we asked OSE to come and settle it, but we had no idea they would put this guy [special master] in our streams to tell us what to do. We just wanted a facilitator, someone with a little authority and outside perspective, to come and try and negotiate something *temporarily* between the ditches and commissioners, but not come in and pretend to rule us all. We didn't think he would stick around for too long."[14] That is, they neither expected nor desired a permanent change in water governance.

AWRM triggered tensions in Gallinas. It certainly did not help state citizen relationships that a special master designated himself "the new sheriff in town."[15] Locks appeared on headgates, and mayordomos lost their ability to flexibly allocate water according to local needs and overflow releases. Parciantes and mayordomos were frustrated during more than one irrigation season after the drought. For a few years after the measures were put into place, many mayordomos jokingly referred to AWRM as the "unintegrated" water resource program.

In the OSE, however, the AWRM program is viewed as a success. The public can actually track that success with a set of maps that show "what percentage" of each of the AWRM basins is ready to be managed for full prior administration. Since most water rights at the individual and ditch level are now dated and mapped, even if the basin as a whole is not fully adjudicated, the OSE can use priority administration. New metering and flow-monitoring devices are a key component of AWRM, and the devices employed can be as simple as a rod with numerical flow measurement numbers exposed or as complicated as housed, protected flow devices in which the data are relayed electronically or logged in a recording system for later download (see figure 6).

There are physical and governance constraints to these forms of metering devices. They affect the actual physics of water flow and influence how, why, and when water release decisions are made. For water users on the Gallinas, even these devices presented new challenges and some new disputes.

FIGURE 6. Typical water flume device employed by the Office of the State Engineer in its Active Water Resource Management program. Photo by the author along the Rio Gallinas, 2010.

The meters installed to monitor flow can disrupt ditch flow, especially narrow, concrete metering devices, like the one depicted in figure 6. Meters too narrow to handle the full spring runoff back up, leading to backflow and consequent flooding in nearby fields. Water backing up behind the meter can lead to slower flow, which affects users downstream if there is not enough hydraulic head in a ditch to cover the entire surface of a field. Antonio Chávez, a small farmer on the Rio Gallinas who mainly grows alfalfa, said he had warned the special master about the problem. "These passive, concrete flumes were just bought in bulk, poorly fitted to the actual ditches and streams, and we told them so. But did they [OSE] listen? Of course not. They just said, 'It will work, and we'll fix it if it doesn't.' What a waste of time and resources . . . it flooded a lot of people back in 2009."[16]

The AWRM program heightened water users' fears about losing local power to administer water allocations. AWRM amendments to the state water code—brought about because of the century-long wait for final adjudications of water rights—gave broader powers to the state engineer and the special masters charged with administering and managing the water allocation process. The Gallinas is still being adjudicated. On the stream system, locals continued to question whether "it was legal" for the state to intervene without a full adjudication. To date, court decisions have favored the OSE.[17] Relationships with the new water master improved after 2012, but the new level of state insertion has been challenging for ditch work.

I heard similar concerns in AWRM basins throughout the state at various times and contexts and spoke with at least twenty-eight people affected by these new measures. Many were concerned about static, inflexible, and perennial state intervention on local streams and watersheds that rarely used local, established expertise. The devaluation of local expertise is unfortunate. As Stanley Crawford wrote, "Thousands of commissioners, mayordomos, farmers and ranchers have been watching rivers and streams for generations, and my feeling is that the engineers listen to them only when they have to."[18] As one Mimbres irrigator put it, the special master has become a kind of "mayordomo of mayordomos and it's really awkward for all of us."[19]

On the other hand, AWRM measures may represent what Tarlock termed one of the "out of the box solutions which involve consensual modifications of existing doctrines," avoiding litigation that might otherwise arise on contentious streams.[20] The program also underlines how little any state engineer actually wants to enforce priority administration. As these meters, monitors, flumes, and special masters deploy across the state, irrigators can also see a new exploratory effort by the OSE: water markets. Since the 1907 water code vested water rights in individuals as private-use rights, New Mexicans have long suspected that water shortages, scarcity, and the treatment of water as a resource might lead to more water transfers and the commoditization of water. One of the first state experiments to explore a live water market was on the Rio Mimbres, in the far southwest corner of New Mexico.

FROM ACTIVE MANAGEMENT TO LIVE WATER MARKETS

The Mimbres is a dry, closed basin. A century ago, it was largely empty country, save for sparse settlements of farmers and ranchers and the miners at the Santa Rita mine near Silver City. Even now permanent settlement seems ephemeral, and it is only the vibrant green ribbon of trees and crops along the Mimbres River that betrays that humans are at work and play in this landscape. At the end of the river's flow, the desert outside the small city of Benson awaits the last trickle. Apart from the higher Gila Mountains, New Mexico's last free-flowing river (the Gila River), and the Mimbres itself, this is a dry gray-and-brown landscape. With no one claiming water downstream, no Indian water rights claims, and a small population, adjudication was completed in 1993 in the Mimbres Basin.

The Mimbres Basin was also an ideal "captive audience," as several irrigators put it, for AWRM (see map 8) because it is a closed basin with no flows going to a major river or ocean. Water shortages regularly occur on the Mimbres. In the past, however, any curtailment on flow was self-imposed. A mayordomo or ditch boss handled the tough call for sharing less water in a dry year. Now, after the implementation of AWRM rules on the Mimbres, a special master would do so.

New Mexicans in semiarid basins recognize that droughts require some unusual interventions if water users along a waterway cannot find solutions. Lucio, a long-time irrigator along the Mimbres I spoke with in 2011, echoed this sentiment:

> No way did we want the engineer coming to tell us what to do. We could see it coming though. They thought that we wouldn't resist. I don't know what drove that whole obsessions with metering and tying to spy on our water use [AWRM], but it seemed like it was headed in a dark direction. We didn't mind a little help to settle water fights or allocation schedules, but we didn't know that whole mess [AWRM] would turn into this permanent thing. We started to feel like guinea pigs. I mean why meter unless you think people are stealing water, or you have another use for that water, right?[21]

Another local irrigator from Mimbres, Alonso, repeated Lucio's concern about perceived water surveillance later in the day. "Now we're watched pretty carefully along this little strip of farms in the middle of nowhere," joked Alonso, as we walked in his backyard. Like other basins in New Mexico, the Mimbres is metered and monitored like never before. Live data feeds and passive water flumes are everywhere in this basin. "I think that people have gotten over the legal thing at this point [adjudication], but we're still learning what comes after, which is the priority thing if they choose to enforce it," Alonso said with a shrug.

We continued for a few steps and then stopped in the shade of a giant cottonwood as Alonso continued:

> I can't speak for all my neighbors, you know, but at this point we kind of feel like we are just a test case, getting poked and prodded, measured, because they really want us to sell or lease our water rights to the cities . . . to them [OSE], it's just an experiment. For us, well, we see it a different way. We see the future, and it doesn't look good for farmers. I feel like at the same time they talk about how important these ditches are to the state, and our culture and everything, well . . . at the same time, they are trying to figure out ways to buy our water from us, or yeah "lease it" [he rolls his eyes], so that Deming can grow more suburbs. But suburbs don't produce food, they just bring more people. And those people . . . well they are going to need more water, right? Where does that end?[22]

The new meters and flumes, to local water users at least, were an ominous sign of things to come.

These meters might seem the barest, crudest, most visible form of the "rule of experts," yet farmers and irrigators, the local experts, increasingly questioned the use and installation of these devices. "This latest episode of spying on us with these meters and stuff is really old news. They've been doing it for years," alleged Tomás, a sixty-three-year-old landowner along the Mimbres. "I remember dealing with the OSE folks in the Deming office in the late 1980s and early 1990s and how insulting, rude, they were at the time." Tomás shook his head at the unpleasant memory. "They would show up asking us questions, but what . . . we couldn't go back in and try to ask things of them? I mean, they were state employees; their job was to settle water issues, right? But every time I took an issue to that office in Deming, they would just ignore me, or make slightly racist comments to the side. Most of the time, they didn't really know what they were doing. Or maybe that's just how I saw it."[23]

On some streams, the presence of a special master has been a net positive, calming interditch disputes or intercultural ones. On others, it has been nearly disastrous as masters have simply padlocked gates and taken the means of governance and control out of the hands of the people who know the stream or ditch best, the mayordomos and parciantes.[24] The OSE is now trying to match local basins under AWRM measures with special masters who are more familiar with the region, unlike some earlier appointments that created friction. What created problems on the Mimbres was the attempt use those meters, flumes, and monitoring devices for a new purpose: marketing water.

Locals were willing to abide by a drought-management plan, the original idea behind AWRM rules. Yet as Doug Freed put it, "We knew it was never just about drought, right? I mean, we could see where this was headed. We were hectored the whole way to sign on to AWRM and to get meters and monitoring instruments on our property and ditches. So once the state engineer said 'Hey, let's try selling your water when it's a really bad drought,' well, that's when we knew there was some other agenda behind this."[25] Clearly, skepticism toward metering "because of drought" and its marketing implications lingers. AWRM started as a drought-related allocation initiative, yet to locals, it seemed to morph into a potential active water *marketing* system.

With no outlet to the Rio Grande, the OSE estimated that third-party effects from water leases and transfers might be minimized. Local water users on the Mimbres did not object to the language or construct of individual water rights as property rights. But the idea of pricing these water rights for lease or sale, to be commoditized and transferrable in a live market auction process, raised immediate objections and challenges on the ditches.[26] Their worst fears about AWRM were confirmed when a live market experiment for water rights was proposed for their basin. The OSE commissioned consultants at Sandia National Laboratories in Albuquerque to explore the implementation of a spot market water-leasing program.

The consultants themselves noted the local skepticism in a 2013 report:

> In two meetings conducted with basin stakeholders a clear indication of disinterest in a market and lack of trust in any project involving water was expressed. Based on this feedback it was decided to continue with the project but without strong stakeholder involvement. The goal of the effort being to develop a template that could be

replicated in other basins in New Mexico. Also, efforts on the Mimbres would not be wasted as stakeholders would likely be interested in the market when a priority call is actually made on the river.[27]

The attempt to implement a kind of universal water market template is clear in these passages—a hope that it could be "replicated in other basins in New Mexico." More important is the first sentence and the clear "disinterest in a market and lack of trust in any project involving water" that might be then rolled into a commodity approach to water.[28] This did not deter the consultants, however, as "it was decided to continue with the project but without strong stakeholder involvement."

Water leasing and marketing could work in the Mimbres. However, even the discussion of a live market sparked fears of water being lost from the land—a slippery slope of rural water disappearing to nearby towns, cities, and larger agro-industrial fields. For locals, then, it was a nonstarter. For experts, it was disappointing. But the OSE and its consultants shrugged off this lack of interest and decided to create a template of a live market, believing that "efforts on the Mimbres would not be wasted as stakeholders would likely be interested in the market when a priority call is actually made on the river."[29]

Some water users on the Mimbres did see possibilities in leasing, although they remained skeptical of outright water rights sales. "All these damn engineer measures parse out water management to silly degrees," Arturo Velarde told me one day as we strolled along the Mimbres River. "I mean, why not just combine the active management stuff [AWRM] with the strategic reserve water so that we could farm most of the time, release water when necessary for endangered fish, or to fight fires and stuff . . . and avoid fights most of the time. I would accept that."[30] Arturo's critique of separate water initiatives like AWRM and water for species addresses how partitioned water has become in the state's view of water sectors. Why not combine them holistically? After all, the AWRM measures are a form of regional integrated water resource management (IWRM), long a buzzword in water resources for creatively and flexibly allocating water between sectors.[31]

Another irrigator admitted, in a guarded voice, that he would take part in it if he could lease an acre-foot or two in years when he was not planting all of his usual crops. "I would do it if it was logged correctly, if I was paid, and it was clear that it was a *temporary arrangement*. I think people freak out if they see this as a slippery slope into water sales, leaving the ditching and going to Deming or some other town down there."[32] There is room for some economic price and lease experimentation here, since there is little economic incentive other than keeping one's water rights to overwater alfalfa or feed crops in this dry, sparsely populated valley.

The state's version of IWRM continues the perceived privilege of outside expertise, and most progressive water managers and planners realize this can be problematic for transparency and stakeholder participation.[33] This has certainly been the case for New Mexico's AWRM program and the strong local reaction regarding this new initiative. Without state intervention, market mechanisms are also being explored for short-term climate and weather-related water scarcity.

Water markets might help avoid future water conflict in many western watersheds.[34] As described earlier, individuals can sell their water right after it has been registered with the state of New Mexico, *even before adjudication* is complete.[35] This has occurred on sections of the Rio Grande River, especially south of the Albuquerque metropolitan area and in the Middle Rio Grande Conservancy District.[36] Such market-driven leases and sales add new complexity to the agency's task of tracking water and water users. The OSE has a hard enough task in adjudication. It will also have to account for water being sold, traded, and leased in specific places and for specific times. These market instruments could be productive if they address the chronic poverty in water-rich places of the state before either large urban or ecological water transfers take place.[37]

As Arturo hinted earlier, there is real promise that such markets or leasing could also produce water for rivers or species, with temporary or seasonal water leases to groups interested in ecology and endangered species.[38] However, trust would have to be built on all sides. As another irrigator, Arturo Gabaldón, put it, "These guys [experts] have to build some local trust here among people first; you can't just tell us what to do without explaining the benefits to us and reassuring people that we won't lose our water rights. Make it clear, make it fair, and then maybe they could do a spot market for water downstream in Benson or for the [silvery] minnows in the Rio Grande if they want it."[39]

The new measures, metrics, and expert practices imposed a new set of consequences. The ways in which expertise was established and carried out in the courts is no longer much disputed. The effects of new metrics and monitoring efforts, however, reverberate across water disputes in New Mexico. Measurements and monitoring created a new visibility for water use between neighbors and water sovereigns self-interested in keeping "their waters" local. Measuring water as a resource did not simplify water use. In many ways, these new quantitative measures complicated access to water and its management. State employees and water traders themselves will additionally have to overcome resistance to pricing and moving water in New Mexico, where multiple water sovereigns define and value water quite differently. Local water ditches have countered some of these new water metrics and strategies in creative ways.

REDEFINING WATER RIGHTS AND DITCH WATER ACCESS

The transformation and production of water into a commodity has been a deeply interdisciplinary pursuit, employing hydrology, law, history, geography, and engineering as supposedly neutral ways to redefine water and to create the new property implications embedded in prior appropriation. A hundred years ago, anyone on an acequia or in a pueblo would have been perplexed at the notion of water rights as private property separate from the ditch itself. Today, not only is this understood, but the same representatives from this communal culture of water have also adopted some notion of an individual water right. New metrics of hydrology were essential to convert local water into state water, into new forms of measurement commensurate with how the state's notion of expertise would function to rule over "its" own water in the future. Claiming that the state owned all the water was a necessary first step in order to give back that water in terms of use rights so that the territory and later state of New Mexico could be acknowledged as the allocator of those water rights.[40] These measures and metrics beyond water duty and acre-feet also continue to have power in the new culture of water experts.

The Gallinas and Mimbres examples given earlier are instructive for other basins. AWRM was implemented in these basins to address drought using new measures and metrics on local waters. Adjudication made water users visible, even in basins like the Gallinas, where the process was not yet complete. Drought brought with it new state powers and state visibility to administer water in both basins. On the Mimbres, these state powers escalated beyond simple monitoring, developing into a dry rehearsal of what a live market for water leasing might resemble. That market approach on the Mimbres never materialized for various reasons but mostly because irrigators were not ready to accept the state's view of water as a commodity.[41] The new water metrics and measurements of the state have succeeded in disrupting, but not overthrowing, traditional measures used by water sovereigns. The new water metrics and measures were clearly fundamental to the rise of experts coproduced alongside adjudication in the twentieth century. After all, experts cannot manage or govern what cannot be measured. Metrics were merely the first step.

Working for the Adjudication-Industrial Complex

What strategies, structures, and silences transform the expert into a spokesperson for what appear as the forces of development, the rules of law, the progress of modernity, or the rationality of capitalism?

—TIMOTHY MITCHELL[1]

I could barely hear the engineers and lawyer in the truck as we bucked over rough washboard dirt roads on our way to the construction site. It was the fall of 2009, and I had asked to take a tour of a new project under construction on the outskirts of Santa Fe, the Buckman Direct Diversion Project. The facility was being built so that Santa Fe could extract its acquired water rights to San Juan-Chama Project water, some 5,200 acre-feet of water per year flowing through the Rio Grande. With this surface water supply, the small city of Santa Fe would then be less reliant on nonrenewable groundwater from the older Buckman wells in the same area (see map 5 in chapter 2 for reference.

As I sat in the back, trying to take notes, listen, and snap photos, the engineer, the Buckman project manager, and a Santa Fe County lawyer chatted. They were discussing the technical challenges and legal particularities of moving water thick with bed load (silt and sand) from the Rio Grande channel, uphill 1,200 feet, to the new state-of-the-art water filtration plant. The plant would end up costing hundreds of millions of dollars. Law and engineering had brought San Juan-Chama water from the Colorado River Basin to the Rio Grande. The finished Aamodt and Abeyta settlements would bring further engineering works not far from Buckman, as discussed in chapter 4. An entire *complex* of water professionals, industry, engineering, and their resulting structures were coproduced by adjudication and settlement.

Back on the road, the project manager pointed. "So you see that giant tank over there, Eric? That's one of several old pump stations for wells currently integrated into the city of Santa Fe water supply, and that will go largely dormant once this is fully operational. We won't have to deplete the aquifer so much with this project water coming in." He slowed the vehicle as we pulled in next to the future filtration plant. "This is going to have everything—reverse osmosis filtration, membranes, carbon filters, and UV treatments for purity. The polish on this water is going to be so good, we'll be able to simply mix it with our other [Santa Fe] watershed sources, and people won't notice any difference." The engineer got out and looked at the project manager. "So, how are you guys going to get the sand and silt out of the [Rio Grande] river water? I mean, that's like half mud, half water." Everyone chuckled, and we continued down to the river.

The streamside coffer dam was being surveyed for the eventual side channel infiltration diversion. The challenges here seemed clear as the Rio Grande was running a thick copper-brown before us. The lawyer chimed in: "How are you separating out the San Juan-Chama Project water from native [Rio Grande] water?" This time, the project manager had an immediate answer: "We will have a whole digital readout for the water sorting and accounting process in the plant, and when we hit our monthly total for water rights given from the [San Juan-Chama] project water, we stop pulling from the river. It's all accounted for." The lawyer nodded, and we drove from the dusty riverside site back up to the filtration plant. The project manager, engineer, and lawyer were all satisfied with this latest arrangement for replumbing regional water while meeting the demands of Santa Fe, state water projects, and legal requirements to track these different streams of water. Here was most of the expert water state, captured in one visit. San Juan-Chama Project water, which had sparked most of the northern adjudications, flowed in the Rio Grande, a portion dedicated to the small city of Santa Fe, and was moving uphill. The era of water projects is far from dead. The cultures of water expertise have kept infrastructure alive well into the twenty-first century.

THE ADJUDICATION-INDUSTRIAL COMPLEX

The adjudication-industrial complex blossomed and profited from the state-mandated adjudications, perhaps even more so from the rising tide of Indian water rights settlements. The early difficulties inherent in the huge task of adjudication were evident early on, and the Office of the State Engineer (OSE) lacked staff and resources from the onset. It is no wonder that the privatization of many aspects of the process appeared in the mid- to late twentieth century.[2]

These pathways of expertise were not just confined to engineering or the courts: history, surveying, geography, cartography, engineering, hydrology, law, and economics have all been woven into the state's pure and applied research to adjudicate water. The mandates of the 1907 water code are remarkably interdisciplinary in their expectations and in their results. Lawyers may suffer the brunt of adjudication jokes, but a wide range of professionals have benefited from the state's failure to adjudicate quickly using mainly public funding and resources. At the same

TABLE 3 Costs of adjudication and settlements in
New Mexico, 2005–2018

Overall adjudication costs for New Mexico:	$65 million
Overall federal contributions to settlements:	$1.5 billion
Overall state contributions to settlements:	$187 million
Total gross cost to public sector agencies:	$1.752 billion

SOURCE: Data compiled from Ridgeley (2010), Richards (2005), and Western States Water Council (2011).

time, public participants in planning groups or meetings may feel their views are not being taken seriously or ignored because of the level of specialized and technical knowledge required to understand single components of the system, much less the whole.

Affixing a number, especially to the private sector aspects, to the costs of adjudication is a challenge.[3] Even a rough estimate of financial costs of adjudication—and the more recent and expensive settlement processes—seems substantial since New Mexico remains one of the poorest states in the United States. Table 3 provides a conservative estimate of the costs of water rights adjudications and settlements since 2005 (a date chosen considering the major settlements with the Navajo and the five Pueblo groups in 2005 and 2006). This rough estimate suggests only the public sector expenditures of these suits, not the private costs or benefits across all fields of expertise.

These estimates included in table 3 also do not reflect any changes to future costs for projects that are underway. Some adjudications and settlements may ultimately cost more than anticipated, as can the water infrastructure associated with settlements. The estimates in table 3 also do not include earlier settlements prior to 2005, such as the 1992 Jicarilla Apache settlement, the 2003 Pecos River settlement, or the 2004 Arizona Water Settlements Act, which may directly affect the Gila River in New Mexico. The Gila may also see another small boondoggle of a dam built for the benefit of very few in the next few years if the OSE and the Interstate Stream Commission continue on their pathway to create another diversion on the last major river flowing wild in New Mexico. These past settlements alone total another $150 million in costs to the state, directly, without federal sponsorship. Only the Jicarilla settlement had federal funds attached to it.[4]

It is difficult to assemble such estimates or understand the cost-benefit ratio, as the costs of adjudication are not publicly available. The state and its agencies do what they can, especially in the Indian water rights settlements, to maximize the benefits to the state and affected Native peoples. This is done, however, without discussing how much past and current litigation has cost all New Mexicans, state agencies, or federal taxpayers, for that matter. But these rough estimates give some clue to the scale and expense of expertise in a poor state. One can only guess as to how much individual citizens have paid attorneys in private transactions during adjudication.

THE NEW SCALE OF WATER EXPERTISE

The scale of today's larger basin adjudications and settlements, involving thousands of defendants and hundreds of lawyers, is troubling. Court hearings, administrative law appeals, and paperwork—mountains of paperwork—are the bulk of what is in adjudication and settlement. Few people want to sit and listen, much less read, these documents. But when taken seriously, the results are fascinating and complicated.

In August 2011, my student Andrew and I sat in on the public, live-action legal editing of the Aamodt settlement language. This was the same meeting mentioned in chapter 4 where attorneys were on edge about the Taos to Pojoaque water transfer. Negotiations would continue for several days, aimed at reconciling 2010 congressional language for funding the settlement with the original locally negotiated agreement from 2006. Line by line, edit by edit, point by point, the latest generation of attorneys agreed or disagreed on the fineries of the legalese on a projected document on the wall. Their heads bowed, chilled by the overaggressive air conditioning, they realized the stakes. In some cases, they were actively rewriting water rights allocations to people in the valley with groundwater wells. There was nothing casual or easy in the exercise.

During a lunch break, Andrew and I spoke with one of the lawyers, Scott, who had been working on the case for much of the past twelve years. If Andrew and I were dazed from a single morning of the process, it was unsurprising that Scott was also weary. However, he also recognized what the case had provided for him professionally—a livelihood. Scott had also put two kids through college on Aamodt, he told us, and the job would not be going away soon. Scott was part of the phalanx of experts and bureaucrats involved in water law, adjudication, and the settlement process. When I asked who else had been involved, he grinned. "You mean, in my generation, or the last, or . . . do you mean the younger associates who will take over from me in a few years? I hope I'm still alive to see this [Aamodt] settlement actually take effect in 2017 but who knows? For sure, there will be some settlement heartburn created . . . and that means younger lawyers will still have a role in the [Pojoaque] Valley long after I'm gone."[5]

The sheer scale, complexity, and specialization of knowledge required for adjudication has given rise to legions of "water experts," from lawyers like Scott to cartographers, historians, engineers, consultants, and researchers. In essence, adjudication produced an entire new industry, an adjudication-industrial complex. Water "expertise" is embedded in the fabric of adjudication. This new complex was publicly and privately funded, bureaucratic, and driven by state needs. Developing and deploying that expertise also took generations. During much of the twentieth century, New Mexico's administrative capacity, its ability to do the science, history, and mapping work demanded in water adjudication statutes, was limited.

Chronically starved of the resources needed to accomplish general stream adjudications, the state enrolled help from the federal government. In a 1908 report, the New Mexico territorial engineer noted the lack of resources and government assistance: "The failure to make appropriation for the expense in General Hydrographic work, by the Legislature, this department was handicapped in this work . . . however, through a cooperative scheme with the United States Geological Survey this handicap has been relieved to a certain extent."[6] Similar statements would be repeated in state engineer reports throughout the twentieth century. State funding did increase, although it never seemed adequate for the massive task. Like most states, New Mexico found itself dependent on federal funding, agencies, and nascent expertise. After all, the first gauges, like at Embudo, were federal. The state water code was also largely based on a federal employee's (Morris Bien) template for western water allocation.[7]

The OSE handled the survey work and state legal suits required for general stream adjudications. However, those undergoing adjudication could not call on the state of New Mexico for help. Defendants—the water users—had to recruit legal and technical help from the private sector. This produced a kind of semiprivatization that devolved work to private industry because of state demands for water rights adjudications.[8] The task was too large and too expensive for New Mexico to both conduct the work required and to offer legal assistance to defendants undergoing adjudication. New water experts would thus be born and coproduced with adjudication.

Generations of lawyers, engineers, historians, OSE technicians, hydrologists, and hydrogeologists provided the necessary direct labor for the state agency. Some experts provided more indirect labor, such as giving expert testimony for the various parties in court. In all cases, however, the bulk of their work for the state was paid for through legislative appropriation for particular projects, rather than by adding permanent staff to the OSE. Adding to the state's public payroll was, and continues to be, unpopular.

Much of this work starts with the new metrics discussed in the previous chapter. *You can't manage what you can't measure* is a common expression in both business and water industry circles, and certainly metrics and measurements of water were fundamental to this new culture of expert water—one in which water itself is considered, measured, monitored, and metered as an object of governance. New Mexico went from a society of local, valley-based, and gravity-fed irrigators to a state where "water experts" and lawyers hash out who will get the rights to that water.

THE MULTIPLE ROLES OF LAW AND LAWYERS

In short, assessing the title to an individual, decreed water right requires the skill of an experienced water attorney.

—ANDREW JONES AND TOM CECH[9]

The necessity of legal expertise is presumed, as in the previous quote taken from an introductory textbook on water management. Like Scott, the lawyer mentioned earlier in this chapter, many lawyers have made lifetime careers from water cases in New Mexico. As another attorney involved in the Aamodt case told me, "Eric, all I had to do was find one big water case, and it's kept me employed for the last twenty-some years . . . remarkable."[10] In the longer adjudications such as Aamodt and Abeyta, three generations of attorneys have worked on the cases, along with legions of historical, engineering, and hydraulic experts.

Many of the local New Mexicans I spoke with have repeatedly argued that there will be only a few "winners" in an increasingly arid western United States, and those will be water lawyers. As author William deBuys summarized, "Water lawyers in the region can look forward to full employment for decades to come."[11]

The lawyers I spoke with estimated that, apart from attorneys employed by the state OSE, some two hundred to three hundred private practice water lawyers have made their livings from ongoing adjudications in New Mexico. Three of these attorneys upped the estimate. Perhaps as many as a thousand, they said, had played some role or made some of their livelihoods from adjudication and settlement in this single state. The cases are complex and important to the attorneys involved. Things can and did get deeply personal, too, as I showed in chapter 2 through some of the Aamodt correspondence between attorneys and defendants when attorneys were asked to step aside.

Law continues to be at the very heart of this industry, and this is no surprise. Any minor or significant change to statute or the water code entails major openings for further legal expertise. For example, a recent measure in the New Mexico Senate would have made protesting new water rights transfers more difficult (SB 665). During a house and senate committee meeting about the contentious measure, a senator joked that it should be called the "2015 Full Employment Act for Water Lawyers." Afterward, another committee member told me that "the senator from that part of the state is always making that joke, because he's a lawyer himself."[12] A jaded local observer had another take: "That's why we're so tired of the lawyers. They're the only ones making money from adjudication or settlement stuff . . . it gets old, but they know they have a lot to gain by dragging these things out or making it all more complicated."[13]

New measures and statutes are written by lawyers in dense legalese, and often only attorneys know what they are saying. Law continues to perform its opaque duty as the technological "software" to the adjudication process in New Mexico and in the other western states. An attorney from Boulder, Colorado, referenced an old joke about his profession: "Eric, everyone hates water lawyers, until people need them on their side for one of these cases. We know it's expensive [pauses, raises eyebrows] ... well, that we are expensive, but if they didn't have counsel, the state or other parties would probably eat their lunch, and they'd end up with fewer rights, or less water in the end."[14]

The seemingly endless legal tinkering, however, may ultimately be of little use. A New Mexican senator put it bluntly in an op-ed piece in the *Santa Fe New Mexican*: "We're concentrating all our energies, and expenditures, on legal solutions—litigating water rights, negotiating water compacts, tediously hammering out water settlements, buying and selling water access—none of which will produce any more real water."[15] The law has produced remarkable clarity on our hypothetical mechanisms for doling out water rights and then enforcing those water rights by date. Yet the law cannot influence precipitation, snowpack, or apparently even dispute resolution. The law coevolved with its principal collaborator, engineering works, as the driver for adjudication. However, it remains unclear whether the law or legal invention or innovation will produce solutions. In some cases, prior appropriation law has exacerbated problems of water scarcity in New Mexico since there is little incentive not to use one's full allocation of water even if the crop does not use or need it.

The law's involvement here does not stop at adjudication or settlement cases. Private groundwater wells, active water resource management (AWRM), and larger concerns about instream flows connected to tribal and endangered species recovery plans are all emerging as (profitable) concerns for legal industries and professions. Maybe the thorniest legal front lies underground.

Groundwater will be the next legal battleground of water use and rights. New Mexico adopted a groundwater code in 1931 and started to include groundwater in adjudications starting in the 1950s. As Scott reflected on these progressive measures of the state, he was more candid. "Our understanding is so legally crude on groundwater ... that when prior appropriation starts getting used by water rights holders and they start putting calls on wells that they think are affecting their surface rights ... boy, it's going to be a mess. Multiply that by a thousand domestic wells going in every year and [shakes head] ... well, you can see the dilemma." As discussed in earlier chapters, groundwater is now conjoined with surface waters in most settlements.

MAPPING EXPERTISE, MAPPING NEW MEXICO

They made a map, and put us on it, whether we wanted to be on it or not.
—LEANN ORTIZ[16]

In 2007 a set of Festschrift publications were released marking the one-hundred-year anniversary of New Mexico's water code. Attorneys and legal scholars made much ado about the survival, flexibility, and current use of the water code. The code mandated that the OSE and its technicians produce a universal, cartographic understanding of water in New Mexico. At the same time, locals typically ignore the code and their junior and senior status unless a drought raises tensions or adjudication sparks rivalries for priority dates.

Field mapping for the state's hydrographic survey was vital to making the water property regime real. Those early maps were "gorgeous, complicated, and often wrong," as Rob Thompson put it. Rob had worked a full year doing field checking in the Taos Valley during the initial hydrographic survey work (1967–1968).

> Those things [maps] were beauties. Imagine a two-foot-by-three-foot or larger linen map, in the end, just works of art really. You would end up with this beautiful work . . . that was . . . okay, difficult to change or edit. So we took our time doing the work on trace and Mylar first, then cross-checking with aerial photography, but it all went to these giant cloth maps. The originals are . . . somewhere in OSE I hope. They were a lot of work, and I'd hate to see those trashed. They deserve to be hung in a museum or on someone's wall as art. They're part of history now.[17]

Figure 7 shows an example of an OSE technical map, hand drawn and meticulous in its detail.[18]

Some of the tract mapping work was also deeply sensitive, since the boundaries between Taos Pueblo and nearby private fields or city properties were zealously watched. Rob recalled the tension on all sides. "Every time we set foot on one of those fields, I felt like I was being watched the whole time by the local farmers, or pueblo folks. It was pretty unnerving but . . . we managed to get along." Taoseños felt like they were being observed by the state; they reciprocated by surveilling the OSE technicians as they did their work.

Cartography remains vital to both local and state efforts to claim water or ascertain water users' rights. In the process of miniaturizing New Mexican waters and lands, reality escapes, and simplification reduces the usefulness of the model. Much on the landscape escaped the state's understanding. There were human errors: mistaken property lines, changes in ownership since decades-old surveys, and crop changes from year to year. Any map, like a census, is a flat and static snapshot of water and water users. One might think that adjudication data from the 1960s, for example, would be updated once in a while, if not regularly. That is not the case. New Mexico, like most western states, does not keep an updated and publicly available inventory of adjudication data.[19]

Geography, and more specifically, applied cartography, allowed for the state of New Mexico to find its own empire of water, map it, and possess it. The cartography of landed and watered property became the state engineer's first tool for historical beneficial use dates. It also had some interesting consequences. Cartography fixed on paper what in real life remains vital, moving, and ever changing in its quantity and quality. These maps created the illusion of state water as stationary, with little recognition that the "state's water" continues to change as it moves downslope and eventually leaves New Mexico. Finally, and perhaps most problematically, most of

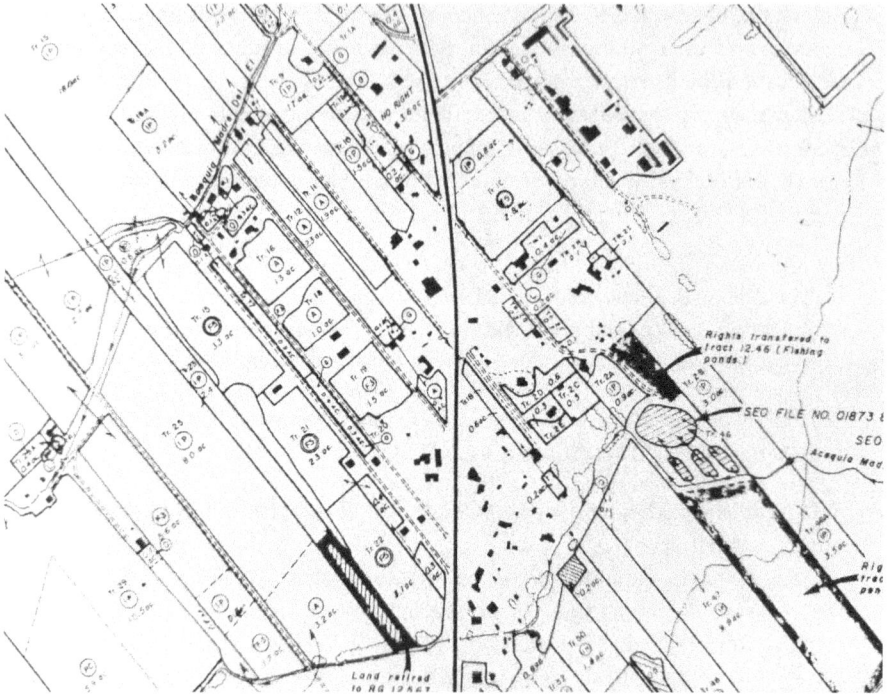

FIGURE 7. Photo of map detail from the Taos Hydrographic Survey work done in the late 1960s. Some updates were added to this map in 1994. Photo by the author. Adapted from: New Mexico Office of the State Engineer 1968–1969.

the specific crop and owner information was obsolete by the time it was finalized. Yet the exercise of the power of cartography is clear in adjudication.[20]

Major rivers are largely predictable in their location, of course, but their quantity always varies and channels may shift. So do the crop choices of farmers and the consumption of cities. The challenge for cartographers was giving some degree of permanence to an always-changing thing. The strategy then became to map the actual land, with crops, and calculate consumptive use of those crops.[21] As Bob Jenkins, a former technician for the OSE, put it, "Some of those early maps of the hydrographic survey as part of the [adjudication] suit, we know, were pretty inaccurate . . . we did the best we could given the equipment, the air photos we had on hand, but it was some of the best data we ever collected on what people were doing with their water at the time . . . since then, well, not much has happened."[22]

Like census takers, OSE technicians like Bob knew that the qualities and quantities they were recording would change over time. The OSE makes no pretensions that their information is current. "We know it's out of date, but those old figures at least give us some idea, some baseline on a conservative measure of how much water was being used, and we did calculate consumptive use, so it's more accurate than a lot of other approaches since that gets to how much water is being used, not just spread over fields back then," said Terry, an OSE technician. Terry had worked across the span of mapping that went from analog paper to the digital revolution and could speak to the vast changes that accompanied the OSE's challenge of modernizing their own maps.[23]

Over lunch in Santa Fe, he nodded knowingly at me as he continued to genuflect on the changes. "It's so much easier now, even if we need people with computer backgrounds, to update that information, but we are usually so short-staffed that no one really looks at the old data to update or account for any changes. We're always tasked with the 'next adjudication,' so the old data rarely get looked at in any systematic way. But any twenty-something working at OSE now can probably accomplish what it took a dozen of us to do in the field, mapping the old way, nowadays. So it's a lot more efficient, that's for sure."

With digital technology and remote sensing, land, water, and irrigation can be detected on the fly and edited digitally on easily updated maps. Even with the new technology, Terry also admitted there was no pretension on the exactitude of the cartographic science of adjudication. At best, the mapping process was a rough landscape guesstimate of how much water was being used across the state. When I asked Terry about these implications of obsolete data and whether this was a problem for estimating how much water was actually used by New Mexicans, he was more philosophical:

> Again, you know, it's a good question, but I think in a way we don't need to worry about it. I mean, yes, water is really, really important . . . not saying it isn't. But just getting that first count, when people decades ago were still farming and using a lot of water, I think that's the thing. It was just to have some data on hand so that you could look at the total, aggregate picture of water, the whole tamale. So I think we're probably okay on water in the state. It's probably a lot better than people imagine it to be, I think. Could be wrong of course, but even bad or average data that could be off is better than not having a clue or flying blind in the water game.

The implications of this obsolete yet generous water-use dataset are fairly tremendous. Even if the static data of water use from previous decades are off by 10 or even 20 percent, the state may have assumed enough water use to provide a buffer for future water use. That is, if water users are not using their full water allocation, there might be more water to go around. Or there may have been at the time these allocations were mapped some forty years ago. Anyone involved in adjudication has long understood the rough imprecision and obsolescence in the process. Mapping, registering archival documents, and the state's ability to monitor and meter have all accelerated greatly since the 1960s. All have been fundamental spatial tools for the legal translation process of water to property in the last century.

ENGINEERING A CONCRETE WATER STATE

If you want reliable water, you have to develop the water resource . . . it won't just magically appear.
—B. SOMME[24]

Engineering flourished alongside state attempts to manage and dole out water. As a former engineer, Bill Ferguson told me, "No doubt we've played an important role in developing the water in this state. Sure, there were some private small ditches before Elephant Butte and Abiquiu Dam, but all that you see here [sweeps hand across a view of the city of Albuquerque] . . . that's because of us, engineers. We did this work. And we're mostly proud of it, even if we have to fix some of the historical dams that need maintenance or redoing once in a while."[25]

Another engineer, Lonnie Davidson, also reflected on her profession and its future.

> Yeah, the era of big dams is probably over, but it doesn't mean that water infrastruc-ture is dead as an industry . . . I mean, look at all the pipelines and regional water systems getting proposed, funded, and built as a result of these adjudications and settlements. Look how expensive it is, too, if people freak out about water quality as well. We are going to be busy for decades . . . and I predict a good chunk of the new waterworks will be about filtration; combating the sediment issues that are every-where in the region; and trying new things with groundwater, brackish water [from groundwater sources], and probably . . . and this will face uphill resistance, you can guess . . . the whole toilet to tap solutions that engineers get excited about, you know, in terms of the challenge. Glad I don't have to sell those solutions to cities, though, because I think at first it will be a tough sell.[26]

This is not cynicism; it is simply realistic.

Throughout the West, engineers are involved in developing a new, less visible hardware of water infrastructure. Colorado, for instance, recently unveiled a first draft of its water plan.[27] Remarkably, the plan calls for an end to megaprojects that transfer western slope water to the thirsty and highly populated Front Range. But the plan does call for "softer" engineering reborn in a format that has less of an impact on environments. Twenty-first-century water infrastructure is more about pipelines and regional connectivity, a less visible form of infrastructure. If the *Cadillac Desert* era of large dams is over, interconnection is the new thrust for most cities and agricultural irrigation systems.[28]

Flood protection and large irrigation projects are no longer the drivers for major water projects in New Mexico. Rather, many of the latest engineered structures built in the state are to use or resolve water rights allocated for municipal water supplies, such as the San Juan-Chama Project that transfers Colorado River Basin water into the Rio Grande watershed. The project provides a mixed water portfolio for the two major cities in northern New Mexico, Albuquerque and Santa Fe. Both cities have created massive projects for taking their "share" or allocation of San Juan-Chama waters for municipal use.

City projects are also less interested in cross-channel dams that have visible and degrading channel effects in the long run. The Albuquerque project uses a half-channel set of inflatable bladders to divert water from the Rio Grande channel but never blocks the river flow completely. Although taken from the Rio Grande, the water is "counted" as San Juan-Chama Project water. In Santa Fe's case, sitting on a small ephemeral tributary, the now-completed Buckman Direct Diversion Project was the answer for balancing water supplies. The Buckman diversion, introduced at the beginning of this chapter, includes a side-channel dam along the Rio Grande and a massive sediment filtration plant (for its location, refer back to map 5 in chapter 2). These are no less simple.

Through a series of pumps, the sediment reduction plant then sends water uphill twelve hundred feet to the Buckman treatment plant, which further filters out sediment and organics. The treated water is then mixed into the other city water supplies and sent through the system. As discussed later, however, these systems are expensive to build and operate. Coffer dams may be less disruptive to the Rio Grande's geomorphology and instream species than traditional across-channel dams, but they have created unintended consequences of their own.

Most recently, engineers have become involved in the state's water management problems through legal settlements. As previously discussed, the Aamodt and Abeyta legal settlements both called for another new regional water system to sort the Indian and non-Indian water claims and needs. This will be yet another water supply project, on top of Buckman, drawing on the already well-tapped Rio Grande and its San Juan-Chama Project water. Clearly, the era of engineering continues but now serves more complicated dimensions tied to legal demands as part of the state's efforts to mete out water rights between its Indian and non-Indian citizens.

Law and engineering went hand in hand with adjudication and the development of western water infrastructure. Dam construction and canal irrigation projects required a legible water rights landscape, driving the need for adjudication. In other ways, engineers may have stopped legal fights because the reservoirs they built hold enough water to allocate a steady flow for downstream water users. Engineering, in some ways, allowed water users to go about their customary ways without enforcing the law. Dams gave state water managers decades of relief from responding to calls on the river that would have been far more frequent, because flow became more predictable if a dam could release a predictable quantity.

Because of the entangled nature of dams and water rights, lawyers and engineers have learned each other's languages. Past governor and former Aamodt case judge Edwin Mechem coined the term *enginawyer* for someone who had mastered the skills of engineer and lawyer combined.[29] Today, any civil engineer will understand the legalese of water lawyers, while lawyers understand engineers' references to water infrastructure mandated (or stopped) because of new litigation or settlement.

HISTORICAL SCHOLARS AND DATING STATE WATERS

History is the ultimate arbiter for who gets better water rights in the west.
—WALT SMITHSON[30]

Like cartography, the uses of archival history also simplify the complicated water histories of New Mexico's cultures. OSE contract historians have to frame their narratives and reports through events recorded on paper and stored in official archives. The archives and documents guide the process of setting priority dates at the expense of oral history. This is problematic in a state where so much of the past was only recalled or recorded in oral history or in oral tradition. This bias toward written sources has long been of local concern. Yet the favoring of archival documents is understandable.

The state makes documentary demands of its own research process. Rumors, hearsay, or misunderstandings enmeshed in oral accounts (only) could easily throw the process into even more turmoil and lead to further conflicts among neighbors. Contract historians working for the OSE are aware that priority dates rest on their ability to find firm archival data and documents. Anthropologists and archaeologists have argued for the use of oral histories and a wider range of material cultural remains in the dating process. However, a date range of first use that might seem specifically narrow to archaeologists would likely be unacceptable to the OSE, which requires exact dating to a year unless those dates fall well before Spanish colonization. Furthermore, archaeological evidence of Native water claims also remains contested since early forms of Native surface water use were done on a seasonal basis rather than as the permanent diversions dated by the OSE. How past landscapes of water use are interpreted are still in play in the courtroom.[31] So is time itself.

Using the archival basis, historians fix prior appropriation dates as per the New Mexican water code. A water user's rights depend on a single date in time, a defining mark of where that user stands in the water line for prior appropriation. This holds true in every western state. Yet that date and a user's rights often only become known and contested in local communities after the state begins adjudication, recruits a historian, and sends an offer of judgment based on that priority date. A use date that may have been murky or simply part of family lore becomes more material. Imagine yourself as an irrigator, dependent on the water you and your family have always received. You start talking to your neighbors who are also under the state's adjudication lens who have gotten offers to see where (and when) your date stacks up relative to others on the ditch. Is your date early or middle of the pack? Or were you or your ancestors too late to this game of history? Will the community still abide by sharing water if you're on a collective ditch? The answers to these questions can affect not only water users' livelihoods but also their relationships with others and how they see themselves.

I spoke with seven current or retired consulting historians who had worked for the OSE or helped ditch members. It became clear that although the historians were respected, their important role in the process was often less visible. After all, people rarely face a historian in the courts unless contesting priority dates or an individual offer of judgment. Most of the historical and archival grunt work by this small sector of experts happens well before any conflict occurs; it largely goes unnoticed. Yet some research has found its way beyond the legal files. As one retired historian told me, "The work we did for the OSE or the private parties involved in these lawsuits did find their way to publication. I mean, a lot of us published articles and books based on all that archival work."[32] Historians' work is also often done remotely. They scour the archives, submit reports, and answer missives on new historical date claims for water. They do not appear in fields like the surveyors or come armed with meters. Historical and archival claims are exchanged on paper, copied, or sent now in email messages and rebuttals.

While the archival research phase of adjudication is often quiet, it is vitally important. Historians played important roles in assessing, understanding, and even refuting prior Spanish and Mexican water principles and in overturning nonexistent water measures like the [Spanish] pueblo rights doctrine.[33] As discussed more in chapter 8, the pueblo rights doctrine was not about the Pueblo Indians but about presumptions made by American lawyers and early state territorial courts that Spanish pueblos (towns and cities) had a prior and superior claim to even agrarian water claims in the state. Historians questioned its validity and ultimately overturned this doctrine, but not before some cities had grabbed water rights on its basis.

Historians have served both the state (for the OSE) and defendants attempting to argue for the ditch-wide dates and communal management principles once common in New Mexico. New Mexicans, as individuals and ditch associations, remain interested in claiming the earliest possible prior-use dates. Their water rights depend on it. However, at the same time, western states are increasingly trying to avoid the use of prior appropriation in the first place.[34] If states are moving away from strict priority appropriation as a doctrine and treating it more like a set of guidelines, one could wonder what this portends for all the work spent on dating and mapping water rights and all the tension the archival process brought to the surface.

THE UNINTENDED PRODUCTS OF
EXPERT STRUGGLES

The experts—those engaged in overseeing and executing adjudications—often admit the folly of the process, or at least the madness of their pace. During a recent 2015 conference between OSE legal staff and the presiding judge (Wechsler), the judge clearly was impatient at the lack of progress. He noted that the OSE's latest efforts to analyze its own annual progress had held the process back even further.[35] The state formed a panel in the early 2000s to explore possible alternatives to the general stream adjudication approach. A few of the ideas from that paper, like creating water courts, gained some traction for a number of years. But the white paper produced little net effect in the end.

As one legal expert working for the OSE told me, "We know it's complicated and expensive, but we've been doing it for so long, that it is hard to honestly say what other system might work better for the state. We've done it for too long to give up now . . . is what I'm saying."[36] Although the alternatives might be more logical, more tangible, and immediate, the state has already invested much time, money, and expertise of all kinds into the current general stream adjudications and settlements. It is hard to see the state and private experts turning away from this system.

The earnest people doing the work on the ground, in the courts, and for the OSE are equally vexed by the time and resources it takes. An OSE adjudicator told me, "It would be great if we had more funding, better-trained attorneys in water law before they come to us, and just . . . more staffing in general. It's a huge state, and I always feel like we're building a perfect replica of New Mexico in a courtroom. There's not enough room, you know, like a diorama of reality where everyone can see these moving parts of dams, laws, water rights, crops . . . it's just . . . [sighs] massive."[37] Water users are not the only ones who are frustrated by the scale and expense of the process.

Adjudication is massive and accordingly generated a massive cadre of experts over the last century to staff it. In this, the state of New Mexico shares much in common with other states.[38] Just as the process was breaking, changing, and forcing new accommodations between water users, so too were the experts navigating these outcomes. In some cases, they may not have realized how their own expertise was reshaping water-user relationships and ecologies across the state. Historical and archival date struggles for beneficial use on ditches also led to tension in these watersheds, even in places that had long shared water (as in the Chupadero case). Engineers did not foresee the massive sediment problems that now plague the Buckman Direct Diversion Project in its struggle to move sand and silt-laden water from the river uphill to Santa Fe.

The increased involvement of private adjudicatory industries is not a *perfectly* neoliberal story of privatizing state functions, yet it did spark a whole set of private enterprises using public monies. Water users need attorneys, and the state requires representation, and often it is the public, as taxpayers, paying the bill for these proceedings. To return to the summer of 2011, Scott the attorney was simply making a livelihood on a matter required by the state. His expertise was necessary and beneficial. The expertise of irrigators along acequias is also important, although sometimes shunted aside in a society that bestows value to technical (scientific) knowledge. The mayordomos are the water experts in their own communities yet are considered "only local" in their knowledge of streams. Because their ditch or watershed knowledge is limited to their segment or valley, they cannot stand in as hired "experts" on how water actually behaves. Yet mayordomos, parciantes, and irrigators on community ditches can often refute "expert" testimony when those specialists take the stand. On more than one occasion, I witnessed an expert give up in the face of local questioning, with a statement akin to "You know more about your ditch than I do!"

At the other end of the spectrum, hydrologists and engineers are viewed as holders of science, technical knowledge, and problem-solving skills. In between are the rest of the experts plying these waters. Historians have conducted work for the state of New Mexico, for private irrigation companies, for citizens, and for the federal government. There is often a recursive dimension to their work; they return to the archives over and over again as water rights defendants contest official historian knowledge and statements done for the OSE. Water lawyers comb through their clients' interests and documents to best represent their claims to water.

The OSE is putatively a neutral party to the entire process. It has employed hundreds of field technicians, attorneys, and staff members, as well as scores of consulting historians, anthropologists, and archaeologists to fulfill its public duties. I am certainly not contesting the use of experts. The process could not happen without experts and some development of expertise. However, the state and its public and private experts have yet to acknowledge their way or knowledge may not be the "best practice" or the only way to manage water. Expert views built into state views of water overlook some particularly useful institutional aspects used by ditch organizations. Nevertheless, as Ruth Meinzen-Dick has reminded us, "there is no panacea for water management."[39] Adjudication itself was dependent on producing water expertise so that the process could happen in legal, archival, and concrete forms.

WATER AS AN ABSTRACTION

Adjudication not only changed on-the-ground relationships between water users. Experts involved in the work created a new, more abstract construction of water. For example, dam operators for the Bureau of Reclamation, the Army Corps of Engineers, and the state often now view rivers and streams as simple numbered boxes in a larger diagram of water flows (see figure 8).

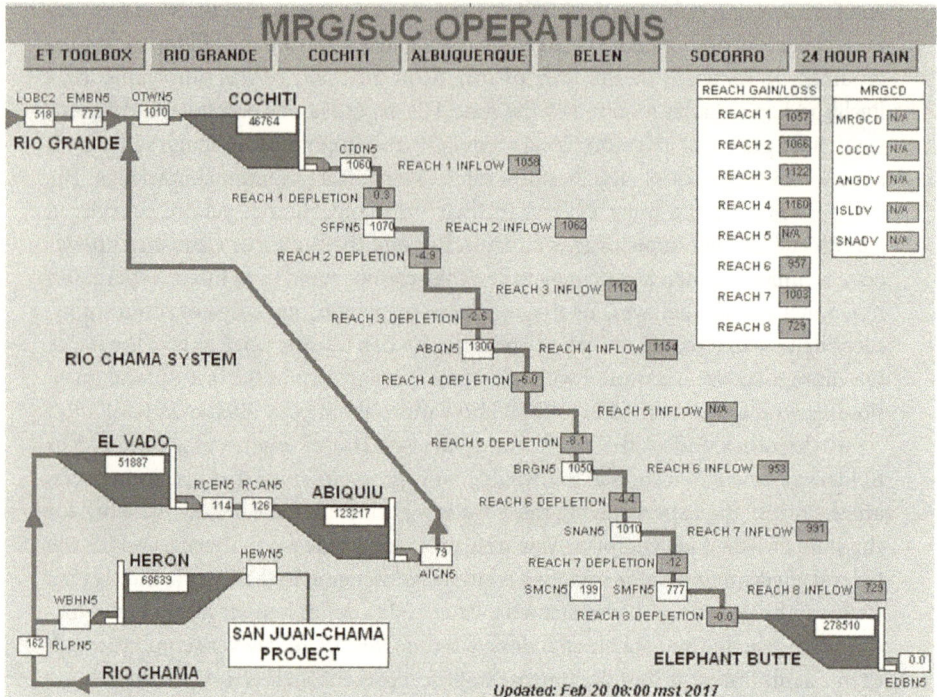

FIGURE 8. One example of expert water: an abstract diagram and screenshot of the San Juan-Chama and Middle Rio Grande waters, flows, and depletions through infrastructure, tributary contributions, uses, and returns to the main stem of the Rio Grande flows. Illustration from the Middle Rio Grande Conservancy District.

Water has been abstracted from a landscape-based tactile relationship, with thousands of users in charge of "their water," to one with several added layers of vertical, hierarchical complexity to its management for the state. Special masters in AWRM, the OSE, individual attorneys representing ditches, consulting engineers, and the federal agencies that manage dams are all now in that bureaucratic pyramid of water abstraction. This is not just about infrastructure but about how water—and the cultures of water—are cleaved by expertise. Expertise and the new abstract views of water have treated water like the state treats identity. The expert view cleaves and separates water by source yet has to manage mixed waters in more complex ways, as seen in figure 8 and in the Buckman diversion example.

Paradoxically, these attempts at quantifying water into some abstract, technical, objective data-based resource have resulted in a heightened sense of regional politics and local distrust over water. Mapping by the OSE during hydrographic surveys encouraged locals to irrigate more. Locals questioned the water duties assigned to maps. Water users questioned historians' use of archival documents and each other's historical water claims, hoping to get better dates for themselves and their ditches. Better data, especially in an adversarial court process, do not depoliticize water. Furthermore, infrastructure projects cemented a more abstract, distant connection to water, as water users no longer had to maintain it as frequently.

New Mexicans pushed back against this form of abstraction, especially the very meaning of what water "is" or "is for" as defined by the state. For example, the parties in Aamodt and Abeyta did not want a state or federal court deciding the terms of water use and sharing in their basins. They turned to settlements and customary water-sharing traditions. In the process, multiple cultural and sovereign inflections on water were made visible and important once again. Adjudication produced unintended consequences, including new alliances, civil society arrangements, and regional water planning.

7

New Water Agents and Actors
in Civil Society

*We can no longer be content with writing only the history of victorious elites,
or with detailing the subjugation of dominated ethnic groups.*
—ERIC WOLF[1]

The elites—the experts—are not the only voices in the adjudication process. As new regimes of measuring and monitoring took hold, civil society groups formed to contest, organize, and question some notions of state and expert water definitions. In some cases, this was in direct response to adjudication. In others, it was because adjudication was taking too long, and the state and water users needed to address new water-planning priorities.

Regional nonprofits that aid small ditch associations have rescaled water-user organization and mobilization in interesting ways. As one of the founders of a water association stated, "We figured we wouldn't let water users be pushed around . . . that we needed to organize as individuals when we weren't happy with how the state was doing its business [of adjudication]."[2] In contesting the state's efforts to regionalize water, these nonprofits matched the state's scaling efforts, enabling ditches to access state funding and political resources.[3] New agents and nonprofits, women in water management, and heightened regional awareness to water planning have also emerged, as have new mechanisms to allow for customary use. Water users are finding ways to achieve more flexibility on the dates assigned to their ditch and canal systems. The agreement that emerged between pueblos and acequias along the Jemez River was one such mechanism.

THE JEMEZ ACCORDS

One notable success in which senior water rights holders came together to resolve conflict during adjudication is the Jemez River accords reached between two pueblos (Jemez and Zia) and five nearby acequias. This agreement was created in the midst of the so-called Abousleman (Jemez River) adjudication in the early 1990s. At the height of frictions, the pueblos requested a temporary restraining order on surface-water use by non-Indians in the basin. A judge denied the request, but more importantly, local leaders saw the madness of fighting their neighbors in court. A new agreement to share water, even during the Jemez adjudication, started to take shape. Peter Pino and Gilbert Sandoval were effective leaders for Zia Pueblo and for the acequia community, respectively. In the 1996 accord, the groups agreed to shortage sharing and that the Pueblo would not exert senior water rights status in a call on the river. Downstream from Jemez Pueblo and closer to Zia Pueblo, the acequias of the region would agree to curtail or halt use in extremely dry periods. In other words, their past relationship would continue.

Observers called it the "first priority administration" plan of its kind, even though it was largely built on the same kinds of arrangements that acequias and pueblos had been doing for centuries. The agreement lessened tensions over water rights seniority and prior appropriation law looming on the horizon while the Jemez River adjudication continues. Observers unfamiliar with the past practices of sharing water touted the agreement as a remarkable "innovation" in addressing water scarcity. State agencies, including the OSE, are also coming around to this customary tradition. Since then, at least three other shortage-sharing agreements have been crafted: along the San Juan in the far northwest part of the state in 2003, along the Chama River, and among a subset of irrigators in the Lower Rio Grande in 2013.

Yet-to-be adjudicated New Mexicans can find important lessons in these accords. One is that local organizing is useful and should not be delayed. A second is that negotiating any desired communal, water-sharing, and territorial aspects of water is vital *prior* to any agreement being settled with the state. The state's water territoriality and its conceptions of water as property continue to prompt water users to craft these alternate arrangements.[4] One cautionary note is that the 1996 Jemez agreement is not a final settlement or an adjudication. It was crafted as a temporary accord so that water allocation and use could continue even while the parties were in court.

Some Jemez Valley residents saw unexpected outcomes in the contentiousness of adjudication, as Vicente, a respected mayordomo on the Jemez, noted in 2010.

> We do actually have problems, of course, and adjudication simply brought the spotlight to some of the things we long ignored with our Indian neighbors, or between parciantes, or even between mayordomos ... but having the state go through this accounting process actually lumped some of that mentality to join together and to make sure we told the engineer that we knew how to work together, that we were actually measuring water, and that most of it goes downstream. The process, I don't

know, it has drawbacks because of . . . it does weird things to people, they get para-noid, nervous, but in the end after all that drama, we figured out ways to use the information and cut a deal with the pueblos upstream, at least for now, to share the water while we sort this all out. People did share their dates, and some documents; we had a good idea of what to expect once the interpersonal water claims were made open [the inter se stage of adjudication, when defendants get to question the other water rights being claimed and awarded by the state].

If anything, we maybe share more information now . . . I mean, I opened an email account because of the damn [adjudication] process just to stay in touch with some of the folks we had to hire or ask questions, like our attorney, some history buff for dates and things. I guess that's good. Otherwise, we tried to keep talking among our-selves during the whole thing. It is hard to keep all that information together. We chipped in and got a cheap computer so that the commissioners on our ditches can keep all that information, the new regulations, updating bylaws . . . sure is easier than using a typewriter, which is what we started out on decades ago. My dad was still using pen and paper for tracking records until the 1950s and all that was in Spanish. We can't ignore . . . we should not ignore that information now, as it can all be used by anyone, anyhow.[5]

His reflections underline an important aspect of these court procedures: they often make implicit agreements more explicit. They resurface old agreements and make them more formalized, in writing rather than as a handshake between water users. As Vicente also noted, sharing data will be a vital part of sharing water on the ditches.

As I write this in 2018, there is still real concern on both sides. The acequias worry about the pueblos' quantified and prioritized rights to the Jemez River. Meanwhile, the pueblos were left reeling by the court's 2017 decision to not award any Aboriginal (Winters) prior and paramount water use.[6] The court, in essence, may follow the historical irrigation pathway, although it has not yet determined the precise nature of Pueblo water rights in the valley. The Abousleman (Jemez) adjudication may provide a firm basis for clarifying future Pueblo water rights claims. This is something that neither Aamodt nor Abeyta managed to produce: a clear legal precedent for awarding quantified rights to Indian water claims.[7] Settle-ments do not have legal teeth, much like the much smaller agreement reached between the acequias and the pueblos in the Jemez Basin. Observers following the Jemez case are particularly interested in any court patterns or agreements that might be useful for upcoming adjudications along the main stem of the Rio Grande. The Jemez case also highlights the regional nature of organizational efforts among water users. Although the affected pueblos, the OSE, and the acequias are still lobbing historical evidence at each other, this allowed for contin-ued shared use of water in the valley. As these suits drag on, new agents and indi-viduals are becoming more active in their claims to secure local waters, even when adjudication has not begun in their valley.

WOMEN IN WATER, NEW AGENTS OF ORGANIZATION

Eric, these people have no idea what is about to hit them, and we are totally unprepared for the adjudication process. If we don't organize and stick together, it's finished.

—TAMARA SIMPSON[8]

As Ellen and I bumped along the washboard dirt road, teeth chattering from the vibration, she expressed worries about arguments on her ditch. She is one of the few mayordomas, female ditch bosses, in New Mexico that I have met over the last decade. While women are now taking more active roles in acequias, Ellen had experienced resistance. Early on, male parciantes often refused to heed her ditch directions. Seemingly, they still did. "People refused to acknowledge me some-times . . . sure, now, they will pretend to shut off their headgate or stop illegally gas pumping off the river if I'm right there in front of them, but if I came back an hour later, the gate's back open again, or the hose is right back in the river, as their little gas engine helps them poach water from the acequia. It's infuriating, really, and it's worse when it's a dry year, and everyone's nerves are frayed." Her frustration was clear, and she had no doubt that her gender contributed to some of the difficulties.[9]

Another mayordoma, Tamara, with whom I spoke in 2011, also acknowledged the challenges of being female in a leadership position long held by community men. Yet she also noted how organizations like the New Mexico Acequia Association (NMAA) were helping fill in leadership gaps left by the passing of the older (largely male) generation. She noted that men under the age of fifty are largely uninterested in serving her local acequia commission anymore:

> It is frustrating, sure. But things are changing, and some of that is just . . . the lack of men willing to do the job anymore, including being a commissioner since it's a head-ache and kind of political, so women are stepping up. We also get some support and some training from beyond the village, so NMAA has been great at that, and the workshops on mayordomo knowledge and training at UNM [organized between the NMAA and Dr. Sylvia Rodriguez at UNM] were really key for me, too, so it's better. But yeah, I get ignored all the time, and all you can do is keep persevering, keep going back to the folks who don't listen until they understand you mean business. Part of it is just emphasizing my place here, driving up and down the ditches, walking parts of it, clearing stuff out, making myself visible to say to them, "Look, I'm in up to my knees in water too, and I'm doing the work that needs to be done, just like you guys did when you were the boss."

Women in water have been active for decades along the ditches and are now part of the fundamental new leadership and organizational efforts locally and regionally. This is as clear on individual ditch association boards as it is on the NMAA board, and it has led to some remarkable collaborations.

FIGURE 9. Tamara, a mayordoma on the upper Santa Barbara River, an unadjudicated watershed that is a tributary to the Rio Grande. Photo by the author, 2011.

Tamara's emphasis on the performance of water and ditch maintenance is a reminder that New Mexicans are not just defending water rights in the courts;[10] they continue to do so in the ditches and streams across the state. At least another dozen women have joined Ellen and Tamara as ditch bosses, as of 2018 (see figure 9).[11] The work has been supported by a variety of groups. Ellen and Tamara both, for example, received some governance training from NMAA facilitators and educators. They also had local experience and knowledge, but as the older men in their watersheds had retired or passed away, much of that knowledge was not transmitted. Thus, the new mayordomo/a training workshops led by NMAA and Sylvia Rodriguez were vital.[12] Such efforts are also bridging what are often cultural divides. For example, the NMAA has organized an annual seed exchange, usually held at Ohkay Owingeh Pueblo (San Juan Pueblo). The event brings together farmers from around the region and focuses on heirloom varieties. People might be coming together, but Tamara still worries what will happen when adjudication eventually arrives in her valley.

Tamara chuckled at the thought of organizing her stretch of the Rio Santa Barbara into a regional acequia association for eventual adjudication.

> It's like herding cats, you know, I mean, people who irrigate up here did what they wanted, for so long, with no one watching but their friends [mayordomos], that everyone is kind of suspicious of doing anything that sounds "regional" since they think the state engineer is automatically involved. He's not, but they have that perception, that suspicion, that something or someone is up to no good . . . or that I'm colluding with the [state] engineer. I've spent hours with some neighbors, some parciantes, trying to show them the merits of why we should organize the acequias into a larger group around here to get state funding to help us with legal costs. Some of them are still . . . they refuse to get it, but most of them see the need. If we don't get some organization going, we'll be eaten alive in courts. We won't be prepared.

Ellen was also worried that the ditches would not come out well in the larger shuffling of groundwaters and surface waters. "Look at what happened between Taos and the Santa Fe area," she said with a wry smile. "Do you really think the water from up there [points to Taos] is connected to down there [points to Santa Fe] in that whole Top of the World farms scheme? I don't think so [shakes her head]. We need to wrap our head around what these failed adjudications do and what we're actually settling for in the end."[13]

New acequia leaders, including female mayordomas and commissioners, are stepping in where men and the young have often lost interest. The lack of information transmission at the local level and on a more informal basis created a void for ditch leadership. A more formal, educational program was needed. Nonprofits such as the NMAA have served an important role in helping local acequias and providing a more regional, cross-cultural platform for education and organizing.

NMAA: ORGANIZING WATER AGENTS

The NMAA developed to help individuals and acequia groups navigate the new financial and logistical demands of the state, especially with regard to adjudication. The NMAA emerged in the late 1980s as acequias began to face attempts by individuals to transfer water rights out of acequia ditches. The NMAA was just one of a number of nonprofits focused on civil resource rights. The NMAA drew on lessons learned and developed by the Taos Valley Acequia Association (TVAA), which was already the most active and visible regional organization.

One of the responses by the NMAA to adjudication and the legal needs of the acequias was to create regional acequia associations designed to match the basin scale of state adjudications. The *Type 1* organizations shown on map 9 *(in lighter shading)* represent the formally recognized regional acequia associations. These organizations have the benefits of aggregating decision-making and funding paths. The member ditches can collaborate and better respond to adjudication. These regional acequia organizations also receive greater recognition within the NMAA's activities and at their annual conference *(congreso)* and more active representation in decision-making and consultations. *Type 2* organizations *(in darker shading on map 9)* are informal groups and tend to have fewer resources with regard to organizing for adjudication. In many cases, they have already been adjudicated.

Membership in the NMAA provides a political access point to state funding, legal support, and educational support, yet not every ditch has joined a regional organization.[14] NMAA also organizes a Congreso de las Acequias, a gathering that draws hundreds of irrigators, mayordomos, commissioners, local politicians, and observers. These events provide valuable networking opportunities, and ditch members can learn from each other, form broader interest coalitions to combat threats to ditch rights, or simply exchange adjudication stories.

The decision to join regional acequia associations is motivated strongly by the size, complexity, and location of the ditches in question. Headwater ditches, if they are situated in an upper watershed with better flows, may have less need to belong to a regional association. Smaller ditches with fewer parciantes have also been reluctant to band together, given their low numbers and getting enough members to participate. Interestingly, acequias are now mirroring the state's approach to adjudication, organizing and responding at basin and subbasin scales. The large scale and number of organizations have put new demands on parciantes, mayordomos, and commissioners.

A century ago, acequias used and governed water only at a single ditch scale, although mayordomos on a shared stream collaborated in early spring to determine how much water would be available. That interditch coordination was traditionally done in early April, and the first Monday of April is now codified for such meetings in the state's water code.[15] Prior to the regional associations, acequias had little interaction outside these irrigation-timing decisions.

Shifts in governance and regulation are also affecting acequias. Acequias were designated as corporate entities able to sue and be sued in 1895. In 1953, they were then made political subunits of the state. In the last fifteen years, the state has also demanded that ditch organizations follow the Open Meetings Act, requiring political and financial transparency.[16] As the state sought to make its waters more visible, it demanded the same of water institutions and ditch organizations like acequias.

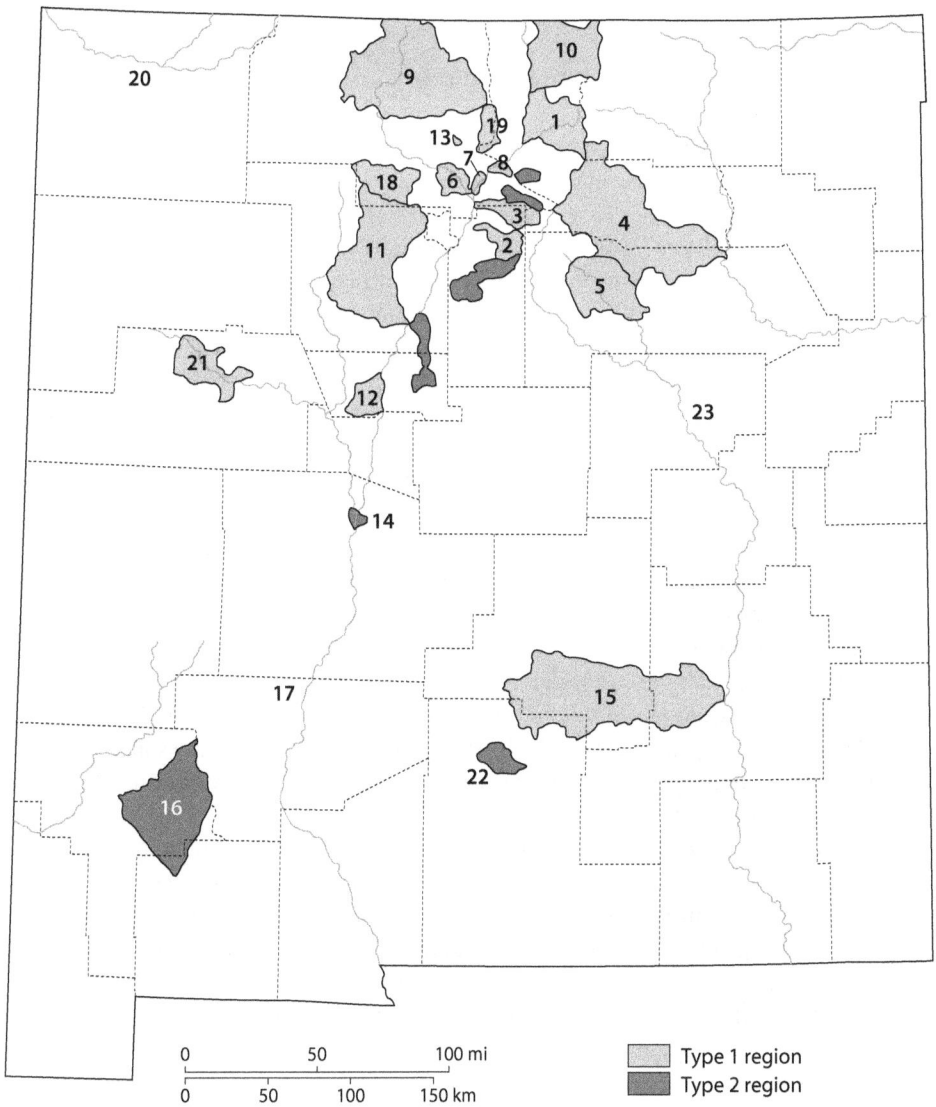

Type 1 region
Type 2 region

0   50   100 mi
0   50   100   150 km

1   Taos Valley Acequia Association
2   Río Pojoaque Acequia and Water Well Association
3   Río Quemado, Río en Medio, Río Frijoes, Río Santa Cruz Acequia Association
4   La Asociacion de las Acequias del Valle de Mora
5   Río de las Gallinas Acequia Association
6   Río de Chama Acequias Association
7   Las Nueve Acequias del Río Grande
8   Embudo Valley Acequia Association
9   La Asociación de las Acequias Nortenas del Río Arriba
10 Questa/Cerro/Costilla
11 Jémez River Basin Coalition of Acequias
12 South Valley Regional Association of Acequias
13 El Rito Acequia Association
14 La Joya
15 Upper Hondo Water Users Association
16 Mimbres Valley
17 Monticello, La Cuchilla, Reserve
18 Acequia organizations have also organized in Gallina
19 Capulin, Ojo Caliente/Río Las Tusas, Truchas areas
20 Community ditches in the northwest part of the state, the San Juan Water Users Association
21 Acequias in the western part of the state, in the Grants and San Fidel area
22 Tularosa Community Ditch
23 Guadalupe County is home to acequias in the villages of Tecolitito, La Loma, Anton Chico, Dilia, and Puerto de Luna

MAP 9. Map of the regional acequia associations in New Mexico. Type 1 regional acequia associations are more formally organized (than Type 2 acequias) for receiving state aid in addressing adjudication. Adapted from the New Mexico Acequia Association.

These legal changes have all produced an uneasy sense of state bureaucratic demands on ditches. State statutes on acequia bylaws created a more standard set of governance templates. Institutional bylaws and templates, however, are not always strictly followed. Some ditches have hundreds of members; others have the bare minimum of three to four members, who may struggle to meet the daily demands of their ditch, plus the increasing administrative requirements.

The NMAA provides governance workshops, with topics such as updating bylaws and meeting the latest state demands for bureaucratic transparency. The NMAA is also working to address a wider variety of water needs beyond the ditches. Most small rural communities without access to a town water utility have formed some version of a mutual domestic water association to provide for household and drinking water needs. Domestic mutual water organizations are vital since very few households take drinking water from ditches anymore.

Regional acequias have also been working to improve partnerships between their ditch associations and mutual domestic water associations, with practical implications for both. When flow becomes low in ditches—through overuse or water seeping away—groundwater reserves are affected downstream, and domestic well associations can suffer. In the summer of 2012, the village of Chupadero (discussed in chapter 2) installed a liner in its ditch to better convey water to the lower valley, hoping to replenish the village's groundwater well for household tap water (see figure 10).

The demands for organized politics and better ditch organization at all scales have intensified for water users. Active and devoted acequia members are often successfully organized, cajoled, and pressured to maintain active roles in the larger regional acequia organizations as well, adding to their workloads. It can all get too demanding. As one parciante from the Jemez River Basin put it, "It's like we're caught in a telescope, and we're either pulled back to our acequia, or the regional organization wants something bigger, something totally different, and our time gets sucked out of the village again."[17] His feelings mirror those of many irrigators and acequia leaders regarding the state's demands for increased visibility, participation, and accountability.

Attending the NMAA meetings can be costly, and not everyone who wants to participate can afford the time and costs of travel. As a Congreso attendee told me in 2009, "I mean, this thing is fun, but we all drove hours to be here, and it just reminds me of a better version of a boring committee or ditch meeting that we already have back on my ditch. So I'm not sure what we gain but simply going to endless meetings. We used to get it all done over this [gestures a handshake motion], you know, not always sitting here and pretending to workshop each other or something."

Parciantes do have internal political and economic access to the state's resources that were unavailable to them thirty years ago. Small ditch and domestic mutual

FIGURE 10. Chupadero residents installing a liner in their ditch so that water from the upper ditches will reach farther downstream and replenish well water near the village of Chupadero. Photo by Eric Shultz. Used with permission.

associations also have better access to legal help through new groups like New Mexico Legal Aid (NMLA). Yet these new civil society nonprofits come with time demands, meetings, and organizational challenges that can sometimes feel like more work. Thirty-eight of my informants mentioned the odd similarities between service for the nonprofit groups and the demands of the state during an adjudication process. Few people want more meetings.[18]

LOCAL AND REGIONAL LEGAL SUPPORT

The NMLA has been a vital nonprofit for parties who typically cannot afford expensive legal counsel, whether for adjudication or for water transfers. Water is just one aspect of the NMLA's purview, but it has become an important component of its practice. "It's hard to believe, but I've been doing this for twenty-five years now in New Mexico," reflected attorney David Benavides, when I ran into him at a meeting in Taos.[19] The NMLA and the NMAA have both been important in representing the "water-rich but cash-poor" acequias across the state and often coordinate efforts when water legislation or broad procedural changes are considered at the OSE.

The NMLA was key in staving off some of the first attempted water transfers away from acequias and remains active on this front. "We haven't always won on the basis of case law," David told me back in 2010, "but often the person trying to transfer their water rights gets tired of dealing with fellow parciantes and just . . . gives up the case or tries to patch things up with folks on the ditch. Social or peer pressure often works, especially if they are related by family on the ditch. . . . But in other places, we have been good at fighting these water transfers away from ditches. It's a real worry."[20] The due process built into water law and state water codes drags out adjudications, but the slow pace also gives water users time to respond and safeguard their interests. Both the NMLA and the NMAA also watch out for any legislation that might speed up questionable water sales and transfers.

For example, in 2015, New Mexico's Senate Bill 665 was under consideration. The bill proposed to accelerate the OSE hearing process for water transfers or sales. The state has officially and explicitly embraced a market-based role for water transfers.[21] Any party can file a complaint to protest water transfers, although they have to provide some evidence of injury, standing, or the fact that their water rights or public welfare would be injured by the sale or transfer. In its original form, Senate Bill 665 was written so that any persons or groups challenging an applicant for a new water right, as a "protestant," would have to provide significant evidence up front proving that they have standing (to protest) *before* they could protest that water transfer.

Proponents of the bill said it would promote efficiency—that the OSE could make faster decisions if it knew up front who was opposing a sale or transfer. An attorney who frequently represented companies and cities spoke in favor of the bill: "It would give my clients some notion as to who opposes it, and why, at the very start, so that we could tell whether it was a valid claim or not. As it is now, we wait for months and sometimes up to two, three years, before the OSE examiner can even give us a response with more information about our applications [for water rights]."[22]

Critics, however, spoke against the bill at a July 2015 meeting of the Water and Natural Resources Committee meeting in Taos. Those opposed saw it as a way to curtail public participation in water hearings, a threat to those who might not be able to provide enough information up front even before the hearing. "What concerns us," said one attorney, "is that this would in effect create a chilling effect for people concerned about water transfers or new water permits, especially if the legislation carried through as phrased since the language specifies that people protesting a transfer could be found culpable for legal fees as carried by the applicant."[23] Others were offended that the proposed bill contained the word *frivolous,* which seemed like judgmental language for due-process civil law matters. Even a senator, who was also an attorney, thought this was ominous language. Citizens, not just politicians, are paying attention to the state's "water language" on all fronts. Senate Bill 665 never made it out of committee, and the NMLA, the NMAA, and other groups continue to push state politicians to recognize due process for ditch associations and acequia norms of water.

DATING BENEFICIAL USE FOR A SINGLE COMMONS

The historical sense involves a perception, not only of the pastness of the past, but of its presence.
—T. S. ELIOT[24]

In water rights adjudications, the past is critical. Neighbors, ditches, and whole sectors move against one another, each trying to leapfrog the others back in time for an older priority date. Any scrap of paper that hints at some early water use is sought out by water users to try to prove the earliest possible beneficial use of water on their properties. People get desperate. I've met irrigators and researchers who are so meticulous and obstinate about interpreting documents that it begins to resemble an obsession. Cities are often in the same difficult position, trying to justify their own acquisitions of water rights through time. Yet it is not just first use in time that matters in the New Mexico water code; it is also beneficial use and proving that that use is ongoing.

Most of the acequias fall between 1600 and the early 1900s as senior water users. One would think that these ditches are fortunate to have more senior dates, giving them some water security in the future. But given that these rights are parsed individually, one of the ongoing challenges is to keep individuals on the ditch acting as a collective. To cope with both internal and external pressures, some acequias and water users have banded together to lobby for single-ditch dates for all water users on an acequia or community ditch.

As part of such efforts, many acequias and community ditches have joined the NMAA in its push for single ditch-wide priority dates. That is, instead of parcel-specific dates for each landholder, single ditch-wide dates would be given, in essence making rights-holding members equal, at least in terms of priority dates. A single ditch date is not always an easy sell for mayordomos and commissioners seeking to gain consensus, especially if some parciantes feel they might be "giving up" a more senior use date. The push for ditch-wide dates reflects long-held communal notions regarding water and can help avoid infighting for dates among parciantes. It is also a form of political-economic resistance to the state, a refusal to be divided and to individuate every parcel of land and use of water. Agglomerative ditch dates reconnect irrigators back to the home institution of the acequia rather than to the state.[25]

Antonio, a commissioner and one-time mayordomo, explained his preference for ditch-wide dates as we walked along the small Rio Santa Barbara.

It's not really practical to have forty-five different first-use dates. You know, it creates chaos, and the mayordomo's job then becomes impossible. It's a bit like we are then responsible for enforcing a state system [prior appropriation] that no one here knows how to use, or . . . no one wants to use it, because it's so difficult. It would make the job of water allocation twice as hard if we had to pay attention to the historical prior-ity date and the amount of water, and what people are planting, all at the same time? It's just an imposition, something we don't see any use for.

Parciantes, commissioners, and mayordomos also had to convince their legal counsels that ditch-wide dates were worth pursuing, since the state water code is set to prioritize individual water users, not communities of water users.[26]

To further keep water local, acequia commissions also established two key new measures: water banking and new water transfer rules to keep water attached to their original lands. These represented a major shift in acequia statutes, and the timing was no coincidence: 2003, right after the state engineer pushed through active water resource management (AWRM) measures (as discussed in chapter 5).

These water-banking countermeasures included into ditch bylaws allow water rights to be temporarily shifted to other users on the ditch but not forfeited by rights holders because of a lack of use.[27] Thus, water can be held within the ditch, keeping water as "local" as possible. Legal protections with water banking allow water rights to stay intact and avoid the threat of forfeiture (back to the state) after five years of nonuse.[28] The water-banking measure cleverly mimicked what larger irrigation districts were doing. For example, the Elephant Butte Irrigation District (EBID) signed an agreement whereby the city of Las Cruces can lease EBID water rights in dry years when fields are left fallow. The Middle Rio Grande Conservancy District (MRGCD) also uses a water-banking model to try to keep water within the district's control. The new acequia water-banking legislation resembles these efforts to avoid the "buy and dry" situation of permanent water sales to cities. Tamara, the mayordoma, echoed the importance of these new measures for man-aging water and water users and retaining what local sovereignty remains. "Just remember, everyone here thinks they are in charge of their own water. We are deeply suspicious when we think that anyone else is trying to manage *us* instead of the water itself." Ditches, in response to state efforts to "reengineer water relation-ships," as Tamara put it so aptly, tinkered with their own ditches and water users.

The slow pace of water adjudications throughout New Mexico also led to some interesting modifications to state water priorities in the late 1980s. For example, New Mexico faced new demands to develop its regional water-planning process in 1987, when a federal court ruled that New Mexico could not block groundwater withdrawals by El Paso, Texas.[29] That federal ruling ordered New Mexico to show it indeed had plans for the groundwater that it was trying to deny to Texas. A state water plan would be needed too, but the regional water plans were in some ways more essential and urgent, given Texas' claims to water that lies beneath both states. A regional water planning initiative started in the 1990s and culminated in the state's first-ever state water plan in 2003, just a year after the severe 2002 drought. The New Mexico Interstate Stream Commission (ISC) was charged with developing and guiding regional water plans. This water-planning effort, I argue, is a by-product of the long and incomplete march toward adjudication and another spin-off effect of its lengthiness.

ADJUDICATION SPAWNS REGIONAL WATER AWARENESS AND PLANNING

New Mexico's OSE and ISC created a new hydraulic map of water regions in which to devise "regional water plans." The New Mexico ISC, essentially a branch of the OSE, is charged with describing the criteria for these regional plans and for their acceptance (both in technical terms and based on public scrutiny). Special concerns of the ISC are interstate compacts, groundwater, and its role in complying with strategic or ecologically mandated reserves, such as the silvery minnow in the Middle Rio Grande River (discussed in chapter 9). The ISC must also maintain flows and water deliveries between compact states, such as Colorado and Texas, on the Rio Grande. For the Pecos River, the ISC only has to work with Texas.

Stakeholders from each area were recruited to comment on, draft, and amend water plans for their areas (see map 10). Such groups are not new. In most parts of the world, there is some version of water governance that bridges the local, regional, and sometimes national scales. In theory, these are laudable efforts to include citizens and multiple stakeholders in more transparent water planning. Like all notions of participatory democracy, however, they are also subject to possible elite capture, or at least cooption.[30]

New Mexicans continue to have concerns about regional water planning and how it may affect or benefit water users. Irrigators wonder why the state is not interested in their *local* ditches, their existing institutions, at the same time they [irrigators] are being asked to participate on a wider, more *regional* scale. This is not unlike the regional shift that both the state OSE and nonprofits have exerted in scaling up demands of local organizations. The regional planning meetings tend to attract the water elite or major local water users. Regional planning is also, technically speaking, a state-mandated effort to have regional watershed groups guide the state of New Mexico's water plan into the future. Yet these plans are merely advisory, and the regional councils have no water governance capacity or power. The process demands local stakeholder time yet does not give them any measurable influence or power in decision-making.

These regional water plans exacerbate further jurisdictional fragmentation and conflated governance boundaries, where overlapping political authorities and microsovereigns do not correspond with the new regional boundaries being used by the state. The separation in regional water-planning regions is largely a matter

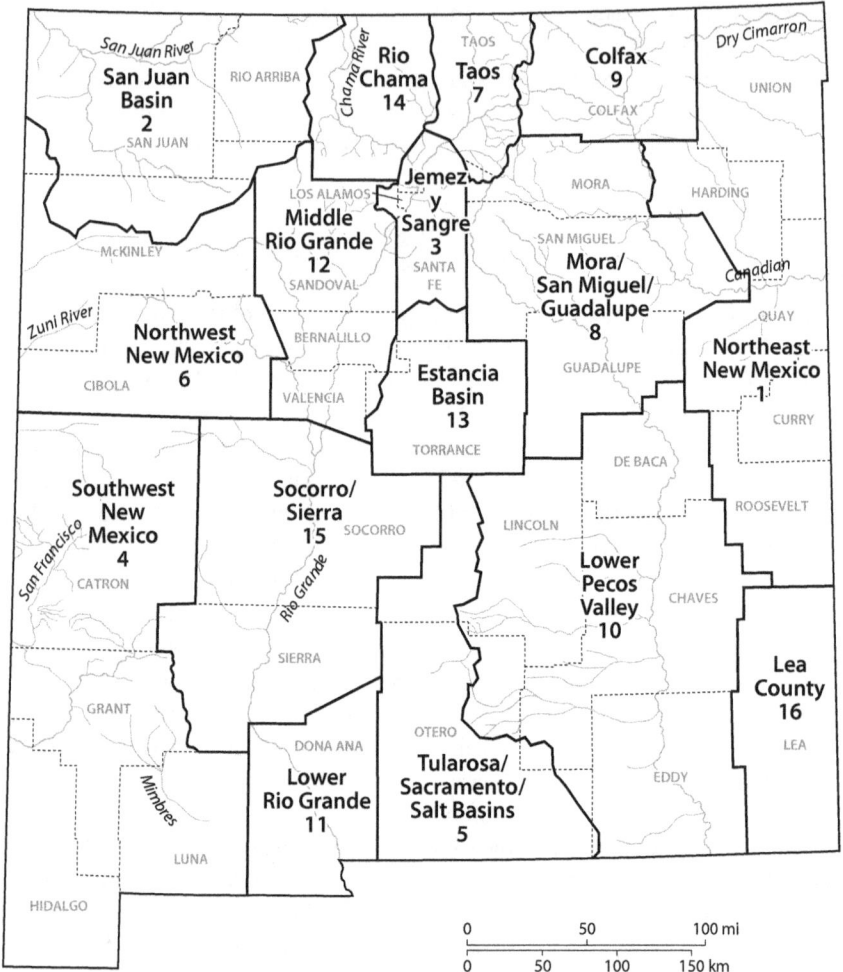

MAP 10. The official Office of the State Engineer/Interstate Stream Commission map of designated water regions for planning purposes, stemming from a 1987 court decision that forced New Mexico to create a more comprehensive, yet regional, approach to state water planning. Adapted from the New Mexico Office of the State Engineer/Interstate Stream Commission.

of geopolitical convenience. For example, the Pecos River originates in region 8 (Mora/San Miguel) on map 10 but continues into the hydrographically named region ten, the Lower Pecos Valley. While the planning regions group large blocks of counties together, they do not correspond to connected watersheds, which can be problematic. Adjudication in some ways rightly uses the basin and subbasin, while the regional water-planning process coarsely relies on waterways and con-joined county boundaries as an organizing principle. Hedging between watershed and county lines for water governance repoliticizes water in a way that may not be productive.

The regional plans are then used, in theory, for the larger "state water plan" assembled by the ISC and the OSE. This provides both an actual and a theoretical role for public participation, even if the vast majority of forum participants are not directly managing water at the landscape level. Attorneys, activist groups, new residents, policy enthusiasts, and interested members of the public all attend. Presenters and panelists speak about the need for "holistic regional planning" and "getting away from our individual gripes" about water management. Abstracting individual concerns to a larger regional concern is one effective way of avoiding difficult topics.[31] For participatory democracy to work effectively on issues of resource governance, all users must feel included or at least considered. Few Native individuals or Hispano members of acequias attend the regional water events, which tend to be focused on apolitical, technical discussions about climate change, urbanization, and moving water to the "highest use for the resource."[32]

As Raul from Taos explained, "Sure, I'm aware of those meetings, and it would be fun to hear about it, maybe learn some new stuff, but some of those require a fee at the door, something like twenty dollars, and I'm not gonna pay that when that's a full dinner for my family."[33] When I pointed out that most of these discussions have no entry fee, Raul simply shrugged and followed with a grudging acknowl-edgment. "Okay, sure then I would. But I'm not sure how that changes anything . . . I mean, they are talking about planning for dams, or conservation, or endangered species, whatever . . . and not about the things we actually care about which is things like date ditches, metering of our ditches and streams, priority dates, that kind of stuff. Plus, a lot of them bad-mouth our irrigation approach of field flood-ing, saying it's old, it's primitive, it's inefficient, all that garbage . . . I don't need to hear that all over again."

Others I spoke with expressed similar scepticism about these regional water-planning efforts. One key person involved with regional water planning expressed frustration with how the state was directing the regions in their planning:

> The state person just shows up to these planner meetings occasionally and was ask-ing them to update their plan . . . and they were baffled. They thought they had a current plan yet had actually never heard back from the state as to whether their original regional water plan was acceptable and in line with other regions. They never received feedback of any kind. Then the state's rep would simply show up years later, telling them to do another plan, to include so and so in their representative bodies, that they had to have a certain level of participation from various industry and citizen groups . . . it was pretty galling to the people who had kept up with the state's water-planning mandate, but they felt completely ignored.[34]

At the state level, a separate nonprofit citizen group organizes an annual New Mexico Water Dialogue to discuss broader sets of watershed region concerns. This group has provided a more open forum for discussing the merits and limitations of the water-planning process and its outcomes. Some critics view regional water planning as only a process—a kind of theater performance where the end is already preordained and well known to New Mexicans who have no power to shape active water governance. One long-time participant in the process of water planning was also honest in her frustrations about how variable the regions' participation can be: "Some regions have it together, have monthly meetings, and already had their plan assembled. Other regional water plans had fallen by the wayside, or gone dormant, if I want to be polite about it."[35]

The regional water-planning process, like adjudication, has illuminated tensions and differences between particular regions and the overall state of New Mexico's plans for water. In my discussions with mayordomos and parciantes, however, it became clear that they considered these regional plans largely symbolic. Some thirty-five interviewees explicitly stated the new regional water-planning efforts to be a consequence of incomplete and long-delayed basin adjudications. They also viewed them as suspect because plans continue to assume that water will be bought, sold, and transferred. The regional water plans also presume that added water infrastructure will be needed. Those plans for new infrastructure and the specter of market-based water transfers that looms in the background both haunt local water users in these settings.

CIVIL SOCIETY CONTESTS "SETTLED" WATER AS A COMMODITY

From the state's perspective, adjudication was and remains the necessary first step to find and account for existing water rights. Built into the water code's assumption is that water-use rights are a private good, which many rural users see as the state's attempt to eventually enable a water market or to more easily allow transfers to new locations. "Every single bit of their [OSE] efforts seems to be geared to get water away from these ditches," Teresa from Embudo told me back in 2011. "Look at what the [state] engineer tried to do on the Mimbres, doing that water market thing. We don't want that here for sure."[36]

Teresa and her neighbors remain nonadjudicated, and they feel unprepared for what is to come. They view contemporary adjudications as simply legal conversion processes to "make our village water a kind of private water, and that's what we object to because then we all lose control over where that water is or goes if people sell out," as Antonio from Las Vegas put it. Their concerns reflect a fear of losing power over water governance.[37] Acequias as institutions have responded to the new demands for increased local and regional political and administrative capacity. Basin-based adjudication has forced a new kind of regional interaction and response from these institutions. For example, the acequias of Rio Arriba County, especially those along the Rio Grande, are in fact putting in their comprehensive water rights claims well in advance of adjudication.

As David Salinas, once active in the Pojoaque Valley adjudication struggles, put it, "We just want these people [experts] to understand that water is for all of us, not just for one of us. It can't be treated like a piece of land . . . that's [land] easier, but water connect us and I feel the engineer and all the lawyers don't get why we are concerned about it . . . if it becomes just a selfish, individual thing, then we're done. There's no point in trying to hold the village or families together. We just get more splintered."[38] The Jemez accord discussed earlier was not an exception, something outside the norm to be called an "alternative management plan." Rather, the informal agreement by basin residents abides by centuries-long customary laws and practices.

The collective stubborn resistance to state efforts to recraft water arrangements and subsurface plumbing is still visible in New Mexico. In late April 2017, defendants from the former Taos suit were once again protesting the planned deep aquifer storage recovery and mitigation wells that are built into the language of the Abeyta settlement terms. They were concerned about the offsets between surface and groundwater, especially near the town of Arroyo Hondo, and what impact that might have on their own domestic wells and ditches.[39] These surface and groundwaters remain unsettled, even in the process of legal settlement. Infrastructure projects linked to these legal arrangements continue to create conflict.

The rise of nonprofit sectors focused on water issues in New Mexico has been remarkable but not surprising. As acequias have become increasingly networked, delocalized, and forced into conversations not bound by the local (only), cooperation among these water-user groups is increasingly common. The bonds between parciantes are changing based on the increasingly secular and institutional needs of acequias.[40] Nonprofits have stepped in to protect the interests of water users in places where the state has instigated processes of mapping and monitoring water use.

As adjudications deployed across the state, nonprofits like the NMAA and the NMLA were key to convincing the New Mexico legislature to provide a legal fund so that non-Indian parties underdoing adjudication could afford some version of legal advice, research, and counsel throughout the process. When adjudication spun out to the now-preferred settlement process, these needs were not diminished. The centrifugal spinning out of adjudication to more informal "agreements" does not mean greater transparency in most cases. In Aamodt and Abeyta, it meant that a small number of people and "parties with standing" got to craft the settlement documents.[41] When New Mexicans feel that their concerns have not been heard by the state or legal representatives, nonprofits are often the only recourse they have.

The scaled and overlapping New Mexican water sovereigns, from acequias, to Native sovereign nations, to the state, and to the federal government, have all bent adjudications and settlements to their own respective needs. The state's efforts to delocalize water created an equal set of regional responses by ditch organizations. In the case of the Jemez adjudication, these were temporary accords to keep water flowing while parties continued to litigate. In others, like Abeyta in the Taos Valley, the settlement included a permanent recognition of customary water sharing. The adjudication practices of the late twentieth century sparked the rise of nonprofits like the NMAA, the TVAA, and NMLA. These new groups, and water planning as an approach, were visible byproducts of incomplete or long-delayed adjudications. This process of civil society response, mobilization, and nonprofit group formation is critical to understand for the story of water use and water users in this state and other western states. Indeed, these new spin-off effects of groups and discussion forums may be the best outcome of unfinished adjudications, as many New Mexicans made clear to me in interviews.[42]

The state's twentieth-century focus on counting small streams, ditches, and upper watersheds in largely rural settings has come at a cost. The OSE ignored most of the largest rivers and populations of the state, as I'll turn to next. Part 3 addresses what remains to be done: addressing the largest cities along the Rio Grande, federal constraints stemming from endangered species, and future water losses under a warming climate.

PART THREE

Adjudicating the Unknown Future of New Mexico's Water

8

City Water, Native Water, and
the Unknown Future

The city [Santa Fe] took the water they needed quietly, kind of sneakily, over the last hundred years. No one really noticed how much was going to be held back behind the dams [Nichols and McClure reservoirs] up in the canyons, and they just went ahead and took it for the city. We knew the acequias were drying up, of course, but we also lived here and wanted to have water into our houses, too, so we were stuck. None of us were really prepared to fight the city since we needed water in our houses too.

—TOMAS ROYBAL[1]

Santa Fe is the capital of New Mexico and home to the Office of the State Engineer (OSE). Ironically, but maybe not surprisingly, it has yet to be fully adjudicated. The basin is still entangled in the long and ongoing Anaya adjudication case. This does not mean that the city hasn't secured its water. Dozens of acequias used to weave through the Santa Fe River watershed, watering fields and orchards. By the end of World War II, most were dry, their water controlled by the city and held in its reservoirs. The water saga of Santa Fe, the self-proclaimed City Different, is in many ways not so different from those in other urban areas of the American West.

Like other cities then, Santa Fe grew at the expense of rural areas downstream. The Santa Fe basin is only the tip of the unquantified water dilemma in New Mexico. The biggest cases are yet to come. Alarmingly, the implicit or explicit water rights of the Rio Grande Pueblos have long been ignored. So has water accounting for the state's largest metropolitan areas, including Albuquerque and Las Cruces. Quantifying Native waters and meeting the growing demands of cities are overdue but daunting tasks.

The personal accounts and experiences of water users in the Santa Fe River Basin hint at what lies ahead for the larger adjudications along the main Rio Grande. They also provide lessons for New Mexico and other western states. The century-old process of adjudication was primarily designed to account for and allocate rural agrarian water. It is imperfectly suited for sorting out the complicated portfolios of urban

145

water needs—even for small cities—in the twenty-first century. As more water is transferred to cities and allocated to Native sovereigns, rural users are often left wondering what actual water will be left in the end.

WATER DISPOSSESSION THROUGH URBANIZATION

The Santa Fe watershed stretches from the Sangre de Cristo Range to the Rio Grande, although only rarely does Santa Fe River water actually reach the Rio Grande. The riverbed is usually dry by the west side of Santa Fe. Towns and villages downstream of the capital, including the old farming districts of Agua Fria and Cieneguilla, see water on an ephemeral basis (refer to map 11). Only in sudden spring thaws of snowpack, or summer downpours during the Southwest's monsoon season, does the river come close to reaching the Rio Grande near the confluence with Cochiti Pueblo.

The ongoing Anaya (Santa Fe) adjudication in many ways exemplifies the gradual yet contested way that cities accumulated water rights at the cost of other water users in the twentieth century, sometimes on legally shaky grounds. Cities big and small face similar challenges as they accumulate water rights. The Santa Fe adjudication also serves as a parable for southwestern urban areas that quickly outgrow their original surface water resources and become hooked on groundwater.[2]

Santa Fe, like some other municipalities in the Southwest, benefited from an assumed—yet historically inaccurate—higher right to water based on historical accident and misinterpretation: a nonexistent doctrine called the "pueblo rights" doctrine putatively based on Spanish laws. Dan Tyler, a historian who served as an expert witness in the adjudication process, has written about the "myth of pueblo rights," as has Peter Reich.[3] *Pueblo* in this case refers to Spanish pueblos—towns and cities—not the Pueblo Indians. The myth asserted that a town (pueblo) has higher rights to water than any other category of water users. City leaders in Las Vegas, New Mexico, asserted the pueblo doctrine throughout much of the nineteenth and twentieth centuries, basing their justification on a California legal practice.[4] This implicit policy of allowing municipal governments to leverage rural water into the cities was used extensively in the twentieth century as both towns grew. It was not until the early twenty-first century that this so-called doctrine was overturned.[5]

In the infamous Martinez case of 2004, the New Mexico Supreme Court finally ruled that no Spanish or Mexican statutes existed in law that allowed for a higher municipal right over other needs like irrigation. This reversed the earlier (1958 Cartright) ruling that had supported the city of Las Vegas' use of this doctrine to accumulate water rights. That reversal embroiled the region in further litigation. The regional acequias around Las Vegas continue to struggle with the delayed and reversed court decision to this day.[6] By 2004, however, the water accumulations by

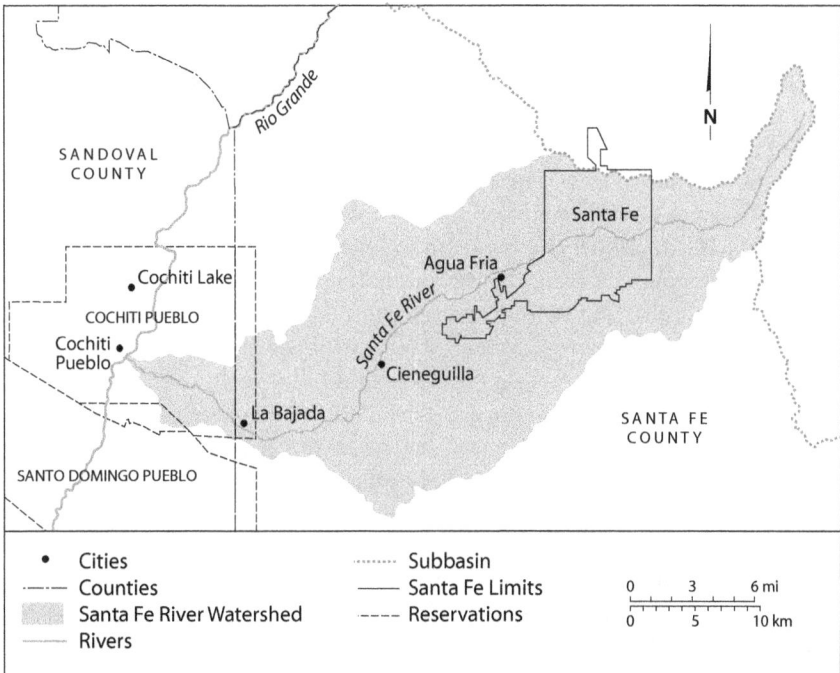

MAP 11. Regional map of the Santa Fe River watershed (Anaya adjudication area). Many of the jurisdictional aspects of the city of Santa Fe and the county of Santa Fe water management spill over into adjoining areas that are not part of the watershed. Adapted from the New Mexico Office of the State Engineer.

Las Vegas and Santa Fe were done deals. Although Santa Fe never asserted a use of the pueblo rights doctrine, the city had effectively usurped the Santa Fe River's flow through reservoirs.[7]

In the end, Santa Fe had physically acquired most of the water in its watershed *without* claiming a pueblo water right by the time the pueblo rights doctrine was overturned. It did so by simply diverting and impounding the water and cutting off ditches lower downstream. The story of Santa Fe's acequias is a clear case of water moving away from the countryside to the city by sheer force of action in the watershed. Yet in 1917 a map produced by the OSE showed thirty-eight functional acequias in the Santa Fe River valley. Even this figure pales to the earlier number of ditches likely in the region (around seventy) before the city effectively piped the water out of a new reservoir in 1881 (see the next section). These 1917 mapped canals irrigated approximately 1,335 acres at the time and largely continued to do so until "about 1940," when many were abandoned as New Mexicans left to fight in

World War II.[8] When veterans returned after the war, much of the water had been appropriated by the city, or little water was coming down the ditches, by many accounts. Seven remaining parciantes and mayordomos I spoke with remembered this time, as well as the city's water accumulation.

The city had control over the river long before World War II. Old Stone Dam, built in 1881 by a private water and irrigation company for municipal purposes, met with local ditch opposition but to no avail. By 1885, a drought triggered major protests and a petition to the territorial governor with 560 signatories from farmers downstream to release water from the dam. The 1885 petition claimed that the "majority of people in Santa Fe have always lived from agriculture producing huge crops, but to this date this has been reduced more and more by being deprived of our water rights by millionaire companies."[9] That petition went unheeded, like so many others to follow, and over the next several decades, the city kept adding reservoir capacity for city purposes. While the initial private irrigation and reservoir would see a short life span, the city would take over efforts to further bring water to the municipality as it grew. Part of the deprivatization was due to the challenges of sedimentation. The reservoir of what was then called City Dam (and later Old Stone Dam) rapidly filled with sediment and was completely filled in a flood in 1904. By 1893, the city was constructing a larger reservoir slightly downstream, Two-Mile Reservoir. The context and expense could not be sustained by private irrigation and reservoir enterprise.

In the early twentieth century, sedimentation plagued even the newer Two-Mile Reservoir, and the city added first the McClure (1926) and then the Nichols (1943) Reservoirs a few miles upstream. McClure was expanded further yet as a reservoir when Two-Mile Reservoir was decommissioned in 1994. With these final pieces of new upstream infrastructure, the Santa Fe River was effectively disconnected from its natural course, and local use of the watershed for irrigation or for seasonal livestock movement was restricted. The river no longer runs through its natural riverbed in the upstream portion.

The city locked down the Santa Fe River basin in 1932, citing concerns that livestock, fire, and human activities in the upper reaches would lead to sedimentation in its reservoir. For city officials, limiting watershed access was a way to protect the city's water storage structures and water supply. Older locals, however, still resent the move, which they link to the livestock exclusions of the mid-twentieth century. This was yet another injustice, a restriction on traditional grazing practices. As David Roybal put it, the watershed restriction was "simply a way to stop my grandparents' sheep from grazing up in the hills like we used to."[10]

Another Santa Fean, Tomás Roybal,[11] had a small plot along the Acequia Madre, the "mother ditch" that still flows through the city's eastern neighborhoods. His uncle served in World War II and left Tomás the land, on which he irrigated less than a half acre of mostly gardens and orchards. He shrugged off any notion that the city had overstepped its water bounds.

I mean, people are always going to get the water first, before some small crops or something, and not many people were keeping animals at home after the war. There just wasn't as much interest after the war. People got into wage stuff and the city grew. I mean, look at it today [the city], and the ditch, well, it is more of a hobby now. It's important and we need to continue protecting what is left of these acequias, but they don't serve the people anymore. No one is really dependent on it. Still, it's heritage, and it should stay, right?[12]

Tomás was realistic about the loss of water and agriculture along the way.

Yeah, Santa Fe was growing, you know. It wasn't a big surprise. To be honest, we understood that people need water, they're going to get it, and there was a little less interest and push for agriculture after the war. I mean we would irrigate, but there was little water coming to the ditches by then, that most of us took jobs that paid money. We stopped farming for ourselves for all sorts of reasons.

Half the people I spoke to about the Santa Fe basin water situation shrugged through a good portion of the conversations. There was a strong sense that "of course" a city would get water if it needed it. I asked if the acceptance was based on the pueblo rights doctrine assertions. Again, most shrugged and claimed they knew little about it but restated that "people should get the water first. We understand." Returning World War II vets also did not pursue farming anymore or simply could not because their ditches had dried up. As Billy Gallegos, whose father served in World War II, said, the world "was bigger, all of a sudden, for these guys returning from the Pacific, and they came back with skills and stuff that had nothing to do with farming anymore. The war changed them."[13] It did not help that no water now flowed through the acequias.

Victor Gurulé's father served in the Pacific, and Victor remembered vividly how his father also returned from the war unable to work the land or clean the ditch:

He had nightmares, yeah, but it was mostly the physical work that was too hard when he got back . . . he went and tried, but it was too difficult to do the back-breaking *saca* [spring cleaning] for the acequia. So he just stopped paying dues to the little lateral, and they cut him off the next year and that was that. We would occasionally keep our fruit trees going with some house water when it got real dry, but we never planted any real crops after that. Before the war, he was still doing corn, some favas, some beans, and chile . . . garlic, the usual food for just family meals. But that war messed him up good, and he wasn't alone. Bunch of other *viejos* [elders] from Santa Fe felt the same way—just picked up and left, or didn't irrigate anymore. The city also built over some old acequias, and they couldn't be used anymore, just paved over, and used as street drainage . . . I mean you see that today, still, old bits of what used to be acequia now just being used as a gutter, collecting trash and stuff. It's pretty sad.[14]

The Anaya adjudication will formalize, when completed, what the city achieved long ago. Begun in 1971, the technical phase of mapping was completed in a matter of a few years (by 1974). By that time, only seven ditches were left, irrigating just over sixty-one acres. Some dramatic changes had taken place along the Santa Fe River watershed. Santa Fe had transformed into a tourist destination by the 1970s. By 1988, when David Snow did his survey of existing acequias for the city of Santa Fe planning department, there were four ditches left. Today, three ditches remain functional, including the well-preserved Acequia Madre, which Tomas still uses for his garden. Two of these ditches serve a kind of yard aesthetic heritage function rather than being critical to food production in any real sense. An ex-mayordomo of one of the Santa Fe area acequias jokingly refers to them as "pet acequias," since few people rely on ditch water anymore for a sense of livelihood. This is no wonder, considering the pace and scale of economic change in the state's capital and how the regional economy is now fully in the sphere of tourism and the military-industrial complex (the latter in nearby Los Alamos). But the parciantes and the mayordomos still service the ditches and remain fiercely proud of the sense of place and history they provide.

Carl, a Santa Fean who used to use water from one of the remaining arterial ditches, saw Santa Fe's water acquisitions as part of a larger issue.

> I mean, what Santa Fe did gradually, maybe with no ill will intended, was take all the local water for itself, bit by bit, as acequias fell into disrepair, and people farmed less, depended less on the acequia. It seemed normal, so unnoticed. Until *poof*, most of the water is in city hands. But this place [Santa Fe] captures what has been happening everywhere in the state, you know, across the West. It's the cities taking more and more water as people leave the fields, the canals, the ditches . . . like no one was paying attention![15]

The city was not attempting to put farmers out of business, but it certainly was active in trying to control its river and the larger Santa Fe watershed and its forests. Impounding water in the upper watershed was a logical next step as the city grew.

Tony, a long-time resident and one-time irrigator off a now defunct acequia, summarized the situation well when we spoke back in 2010.

> No one talks about Anaya . . . it's amazing. The city [of Santa Fe] did as they pleased because well . . . I mean it was the state capital, and even people who have senior rights on the acequias are not going to stop water from getting to people. I mean, that's a built-in principle, you get water to people . . . so you could say the crops and orchards came second. On top of that, a lot of long-time Santa Feans moved away, especially after World War II, or could not afford to farm or got a job at the lab or downtown in a hotel or something [shrugs]. That's how it went. I think the city was also pretty clever in using that whole pueblo water doctrine, not the Indian thing of course [chuckles] . . . I mean that weird idea that cities could use and take all the water they needed for its own citizens even at the expense of other water users.[16]

Historical injustices still stung for many of the informants I spoke with. Tony was well aware of the pueblo water doctrine, the long-abused and nonexistent clause in Spanish water law that putatively gave cities priority in their water needs:

> That was pretty sneaky. But they [Santa Fe water managers] weren't alone. It was sort of an old Spanish vagary in the existing water principles, and I think Las Vegas [New Mexico] used that rule as well . . . to just get more water, as much as they needed. But it's not really a shock to see all these old acequias sitting dry. The city took what it needed, agriculture became less important . . . acequias are now just . . . I don't know, like memorials or commemorative things, not really functional. Sometimes they're still there, but people don't know it because the city or NMDOT just turned them into drainage ditches near roads. That's pretty typical everywhere . . . but especially near the big town and cities where people left farming a long time ago.

Overall, Anaya did not lead to conflict among neighbors or scrambles for priority dates. Those disputes took place, of course, as I discussed earlier. But by the time the lawsuit was filed in 1971, much of the water was already held by the city. Despite its regional exceptionalism and architectural preservation efforts, Santa Fe's water history is typical of other cities in the American Southwest. The city's quiet and incremental approach is hidden from most visitors and tourists today, but not to those who experienced it.

A CITY NOT SO DIFFERENT

Hike through a little nature conservancy open space just east of downtown Santa Fe, and you can understand Santa Fe's historical and contemporary water challenges. Today, the remnants of these two older dams (Old Stone and Two-Mile) are scenic wetlands within a small nature preserve, a joint venture between the Audubon Society and the Nature Conservancy. One can still find the old rock-rimmed small dam (Old Stone Dam) in the middle of this open space, often trickling with some water spilling over the rim, as if the dam remembers its old function. From the interpretive signage describing the old reservoirs, to the trailside trees gnawed on and worked over by beavers, a visitor gets the clear sense of the shifting scale of water dependency for the city before it started dipping into groundwater.

Visitors to these older reservoir sites do not have to be dam engineers to see the challenges posed here for storing water. There is little ground cover on the steep hillslopes to hold back the crumbly sediment and gravel that surround the old Two-Mile Dam. One large metal panel wraps around the dam site, put in to hold back the sediment with little success. The two larger upstream reservoirs, Nichols and McClure dams, still provide 30–40 percent of the city's water needs in any given year. The rest comes from a variety of city wells and more recently, the Buckman Direct Diversion Project that takes Santa Fe's share of San Juan-Chama water out of the Rio Grande. These concrete structures of the city's water past, however, are silent about the social consequences to local Santa Feans.

The hard work of impounding and urbanizing water along the Santa Fe River watershed was a fait accompli. Yet local politicians in and around the Anaya lawsuit remain restless about the lack of a final decree for the Santa Fe area. As of 2017, a new effort to push along the Anaya adjudication was underway in the form of a state senate memorial. While these memorials do not have the power of an act, the sponsoring state senator (Elizabeth Stefanics) actively encouraged the negotiation process begun in 2009 by the Interstate Stream Commission (ISC) on the Santa Fe River. As of early 2018, no formal response by the OSE or the ISC had been issued regarding the Anaya adjudication or the symbolic shove administered by the senate memorial in question. Small farming villages downstream of Santa Fe hope that the negotiations might allow for some Santa Fe River water to make its way to them.[17]

Santa Fe serves as a microcosm for the water challenges faced by cities in New Mexico. To the south, Albuquerque, the largest population center and economic force in the state, has bided its time. Albuquerque and its water authority, bedroom suburbs like Rio Rancho, individual farmers in the larger Middle Rio Grande Conservancy District (MRGCD), the six Indian Pueblos, remnant acequias within the conservancy's boundaries, and downstream users all have a lot at stake.

THE BELLY OF THE BEAST: THE MIDDLE RIO GRANDE

The Rio Grande main stem is the proverbial eight-hundred-pound gorilla in New Mexico's water adjudication process. The bulk of Pueblo waters that lie on or around the river have not been quantified.[18] The issues are not just about people. Since the 1990s, the endangered silvery minnow has taken its own cut of water rights from the river's flow (as discussed in the next chapter). Town and country, human and nonhuman, Indian and non-Indian—all of these binary water constructs are found along the state's largest river. What awaits the OSE is the ultimate test of the legal, administrative, and political process. The population of greater Albuquerque is nearing a million as municipalities and suburbs are clamoring to find "new" water, and farmers are worried they will be left dry. One assessment of the MRGCD was dire: "Some experts feel we are headed for a train wreck because the MRG [Middle Rio Grande] is over-committed with regard to water rights, meaning there are more claimed water rights than actual wet water in most years."[19] The situation is further complicated by the expertise and infrastructure that have been put in place alongside or because of adjudications.

Like Santa Fe, Albuquerque struggled to keep its waters clear of sediment from flooding along the Rio Grande. In the 1880s, southern Colorado irrigators in the San Luis Valley started irrigating on a more intensive basis, and the Middle Rio

Grande near Albuquerque filled with sediment and experienced increased flood-ing. By the early 1900s, the Albuquerque stretch of the Rio Grande was clogged from bank to bank with sand, silt, and flooded riparian habitat. Photos attest to this real issue, as do documents describing the creation of the MRGCD. Although acequias remained functional despite the sediment problem, larger needs for flood control, municipal water, and irrigation prompted the development of the MRGCD in 1923. Conservancy districts have far greater powers than small community ditches, or even federal irrigation districts, in that they can self-govern and tax land owners as their own distinct entity.

The MRGCD included the majority of irrigated land near Albuquerque, stretch-ing between Cochiti Dam in the north to the small town of San Acacia in the south (see map 12). Part and parcel of the MRGCD was the construction of El Vado Dam, completed in 1935 on the Chama River and now run by the Bureau of Recla-mation.[20] El Vado Dam was built for flood and sediment control and to store and allocate the conservancy's agricultural water during the dry season. El Vado Dam and the later Cochiti Dam (1973) now serve the additional purpose of providing water to the six pueblos along the Rio Grande: Cochiti, Kewa (Santo Domingo), San Felipe, Santa Ana, Sandia, and Isleta. These pueblos have defined federally reserved water rights.[21] The MRGCD was awarded nearly a quarter of all San Juan-Chama Project water, and Albuquerque was awarded more than half the imported water. Here on the Middle Rio Grande, the lion's share of Colorado River Basin water is split between the largest conservancy district and the largest city of New Mexico.

The conservancy district's creation (in 1923) was yet another instance in which control over water allocation and the scale of control shifted. No longer did local parciantes and mayordomos have full control of when to open the headgates, how to distribute water, and when to levy delinquency fees. These controls were wielded by the so-called ditch riders working for the conservancy district. The conserv-ancy district also had the power to tax. Many parciantes could not afford these new special taxes, but unlike acequia dues, the MRGCD's taxes could not be negotiated or delayed. The loss of sovereignty and new taxes did not go uncontested. Dozens of existing acequia systems and their users fought the creation, implementation, and operations of the MRGCD. The conservancy district ultimately prevailed. This dispossessed former acequias not only of their direct access to water but also of their governance powers and flexibility in financial arrangements. Many farmers lost property because of their inability to meet the newly imposed conservancy fees and taxes. The lessons of the MRGCD, and the resulting property disposses-sion along acequias, were part of the reason Taoseños defeated the proposed Indian Camp Dam conservancy district (chapter 3).[22]

The new conservancy district disallowed diversions by the acequias within its boundaries other than some minor allocations to specific lateral ditches during

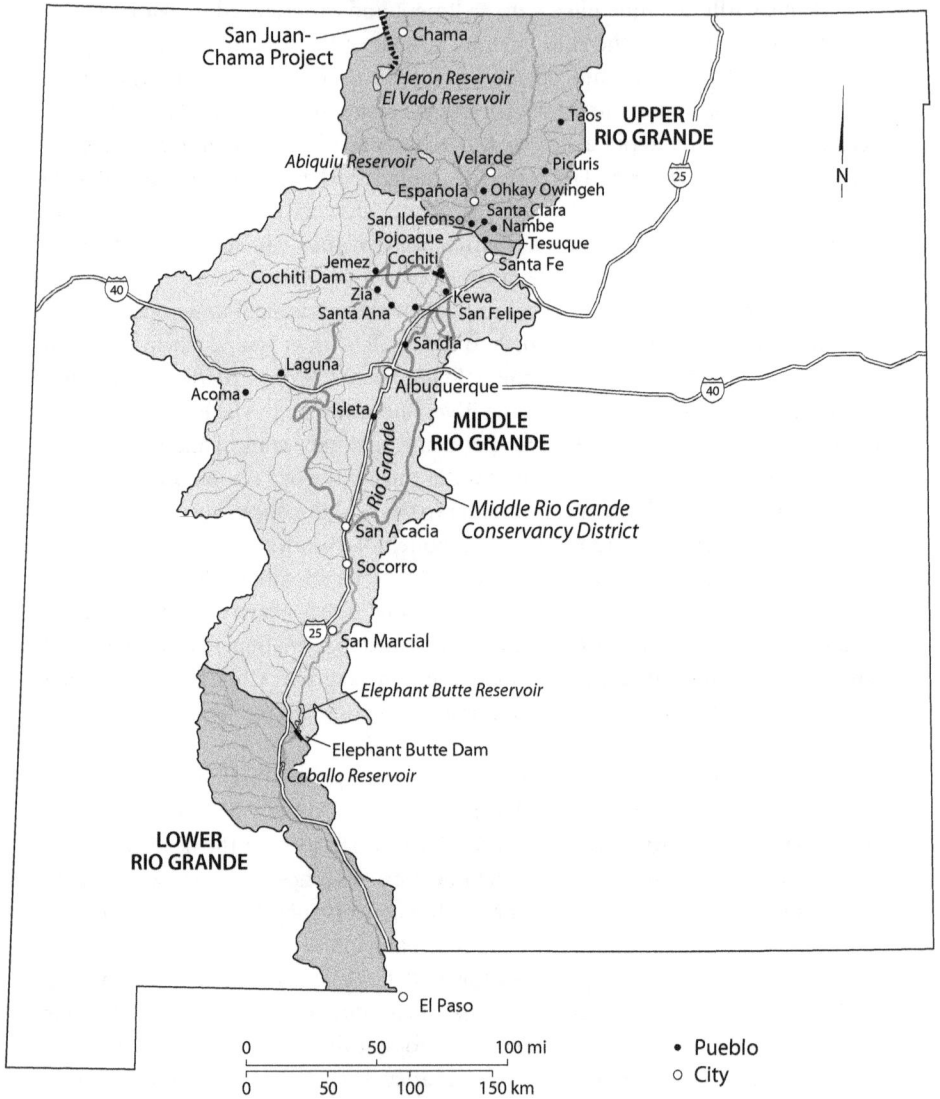

MAP 12. Location and context of the Middle Rio Grande Conservancy District (smaller bounded area) within the larger Middle Rio Grande stretch and major works on the Rio Grande. Six of the Middle Rio Grande pueblos have recognized federal "prior and paramount" water rights. All other Native water rights await adjudication or settlement. Adapted from New Mexico Office of the State Engineer and Middle Rio Grande Conservancy District maps.

irrigation season. The MRGCD accumulated water rights and taxation authority via legislation and represents another kind of legal violence that moved water authority away from local water sovereigns like acequias to a larger jurisdictional office interlocked with other developing water agencies.[23] About seventy Pueblo and Hispano acequias were incorporated into the MRGCD. The loss of acequia water governance and autonomy, similar to Santa Fe's urban acequia experience, happened between the 1920s and accelerated during the 1940s.[24] Interestingly, just ten years ago, a new set of acequias in the south valley of Albuquerque reactivated their institutional base to assert local ditch control once again. They have organized as the South Valley Acequia Association. How these new or reformed acequias will fare during or after an adjudication of the MRGCD area is unclear. But they do now receive water from the main laterals, to then distribute among themselves with an elected mayordomo.

GROWTH WITHOUT A PLAN, OR PLANNING FOR GROWTH?

Look, there is no central policy or plan on water. Every damn state does as they please. The feds manage most of the dams, the state pretends to manage water, and everyone else pretends to own the water . . . when they don't. We're simply fumbling through this all . . . all of it.

—MARIO SANTIZ[25]

The new scope of the MRGCD's power and its separate status complicated New Mexico's jurisdictional map. The MRGCD is the largest conservancy district in the state, extending 125 miles north to south, consisting of seventy thousand acres of farmland, and serving some eleven thousand irrigators, with *presumed* water rights to some four hundred thousand acre-feet per year.[26] Residents within the conservancy district's boundaries do benefit from a more dependable water supply. However, that water is not yet adjudicated and likely already fully allocated, if not overallocated. In 2003, the Albuquerque Bernalillo County Water Utility Authority (ABCWUA) was formally created for the city's drinking and wastewater management system. Together, the ABCWUA and the MRGCD account for a massive share of water rights along the Rio Grande.

In developing the MRGCD, water was treated both as a flooding risk and as a fuel for growth.[27] Over time, however, the conservation aspect of the conservancy district has been highlighted, since the MRGCD also manages thirty thousand acres of forested riparian cottonwood galleries. These riparian spaces and trails have gained new value since the 1970s as recreation became a key urban amenity for Albuquerque residents. Because the district was created with more than just irrigation in mind, recreation is easily accommodated in the management plans.

"We're simply fumbling through this all," Mario Santiz said in an earlier quote, a sentiment that some other water managers shared. Mr. Santiz used to be a city water employee for Albuquerque, and the nexus of fragmented water policy, non-planning, and bureaucracy frustrated him until he quit in 2006. Water managers face multiscalar challenges, and Santiz was openly concerned about who would manage what specific waters, where, and when. Cities have conserved water by actively pricing water at more aggressive rates and by passively relying on US Environmental Protection Agency policies that lowered minimum flow rates for toilets and showerheads. There is less innate resistance to paying for water in an urban context, even if water is stubbornly difficult to commoditize, much less privatize.[28] However, the total aggregate water demand still remains stuck on a high plateau of consumption.

However, the century-long delay in adjudication has not kept Albuquerque and its suburbs from expanding their water footprints. In the MRGCD, those who actually farm near the city are concerned: "They're sprouting the suburbs, and we just know this will come back to haunt us when we finally get adjudicated because the only water left to divvy up and dole out is going to be this water, the conservancy's water. Where will that leave us?"[29] The larger concern is that less water will be available to meet both the old known and the new unknown demands, whether rural, urban, or ecological. In the latest episode of suburban water struggles, a proposed development named Santolina is being considered on the southwest side of Albuquerque. Immediately, the proposal raised concerns among water users already worried about limited water sources in the MRGCD. The developers have tentative approval, yet public hearings continue to be held, with major sticking points being where the water will come from.[30] Irrigators in the MRGCD suspect it may come from their conservancy district.

The six Middle Rio Grande Indian Pueblo groups must also be allocated their water rights upstream and downstream of Albuquerque on the Rio Grande, and this will vex state, federal, and conservancy representatives. Only when Native rights to water are accounted for will the OSE, the MRGCD, and the city of Albuquerque have a real idea of what actual water exists. Currently, the conservancy and the city have only presumed paper rights to water. From the Pueblos' perspective, it is unjust that wet water was acquired by cities and suburbs over the last century before Pueblo rights were quantified and allocated. Furthermore, the Pueblo have often paid a steep price for the districts' flood control and irrigation needs.

NATIVE PEOPLES, NATIVE WATERS

We were here long before anyone else, we had respect for this water, and it still shows today. Look how people mistreat the water. We have to find a way to give back to the water . . . everyone here in this meeting has been talking about water as a resource, something to take, to use, but when do we give back to the water?

—ACOMA PUEBLO ELDER[31]

The Middle Rio Grande Indian pueblos along the main stem of the middle and southern Rio Grande have been unaddressed and ignored by the official adjudication process or settlement endeavors. Cochiti, Santa Ana, Kewa (Santo Domingo), San Felipe (two of the largest stakeholders), Sandia, and Isleta Pueblos all await formal adjudication or settlement of their water rights. New Mexico's sovereign tribal entities are located along the main stem of the Rio Grande, as well as in drier watersheds in the west and northwest parts of the state (map 12).

As discussed in the previous chapter, the Jemez adjudication proceedings may set a new water standard for quantifying the remaining pueblos on the main stem of the Rio Grande. In the interim, the six Rio Grande pueblos at least have a "prior and paramount" water rights agreement in 1928 from the MRGCD. These are not formal settlements, and future claims may be significantly greater than the acre-feet currently agreed to between the conservancy district and these Pueblo groups. By the current arrangement, these prior and paramount acres (about 8,800) receive water first, before all other users in the conservancy district. Those waters are typically delivered to the six pueblos via the main native flow of the Rio Grande, but they also get served from El Vado (Chama River) water when necessary in times of scarcity.

As the quote from the Acoma elder makes clear, Indian water rights will be major hurdles for any future water allocation, or reallocation along New Mexico's major river. The tribes may also be able to leverage some eventual allocation to environmental purposes, hinted at in the Acoma elder's statement about "giving something back to the rivers," should they choose to put their prior and paramount rights back into the Rio Grande. And yet Acoma is not part of the MRG and sits far to the west. They control few water rights and are thus on the outside looking in, as Acoma's rights also remain unquantified. While unlikely in the short term, since the Rio Grande pueblos' water rights have not been awarded yet, putting some future water back into the Rio Grande is a distinct possibility. The six pueblos of the MRG will have to determine all this themselves once their waters have been sorted and awarded.

One of the more painful water infrastructure episodes along the Rio Grande was suffered by Cochiti Pueblo in the decade from 1965 to 1975, as the Cochiti Dam and Reservoir were installed midstream, right above their inhabited and centuries-old pueblo (see map 12). The perceived need for flood protection and additional water storage in Cochiti (Reservoir) on the main stem of the Rio Grande overrode the pueblo's resistance but did not lessen the pain.[32] Regis Pecos (2007), a former governor of Cochiti Pueblo, elaborated on how traumatic that concession was at the time:

As the elders conceded and succumbed to the incredible political pressure of condemnation of the land and construction of the lake, it was painful to see the elders, members of our community, full of emotion, full of tears, and reflections of a sense of helplessness in their eyes and faces. They spoke with a deep sense of hurt that they had failed as the stewards and the protectors of this incredible, beautiful, and sacred place to our Pueblo people. It was the heart of what gave meaning to our lives. It was very painful for me to witness this helplessness. All my life I had seen these same men with a sense of wisdom and vision, strong spirited, always acting with a sense of certainty and assuredness. Now they were reduced to this helplessness. I had never witnessed such hopelessness. It was frightening. The future seemed uncertain. This was one of the most tragic episodes in recent history for the people of Cochiti.[33]

Largely built for additional flood control and to equalize flows to the MRGCD's irrigated lands, the dam still stands and is now vital for timed water releases in plans to manage the irrigated growing season. This is small comfort for the people who live on the northern edge of the conservancy and whose lands were often saturated until the late 1990s because of chronic seepage from Cochiti Dam. The Cochiti "heartlands" were inundated to provide for Albuquerque's flood control, suburban development, and as one current resident of Cochiti put it, "the white man's agricultural plumbing."[34] Gerald, a respected member of the Cochiti in his fifties, reflected on these changes:

> Maybe the worst part of this whole episode was that they named the dam after us, I mean . . . Cochiti Dam. We didn't want the thing in the first place. A lot of the elders, including my father back then, were really opposed to it, but we didn't have any power back then. They [the feds] kept pushing us to accept particular deals, tried to buy the Pueblo off with some perks and benefits, but I think it's the name that bugs me the most today. It is like putting up a really offensive statue in your backyard that you don't want, and then they name that gross thing after you, even though you hate it, didn't want it, think it's a desecration. That's my take on it. . . . If only we had been able to delay the first motions on construction, we might have been able to stop the dam, because it's a lot harder now to build them, you get it? If we had better lawyers back then, like we have access to now, it could have been maybe stopped before it became what it is today.[35]

Cochiti was one of the last large dam projects in the state, finished in 1973, near the tail end of the golden era of dam development.[36] The results of the dam project did have some benefits for nearby residents. In the process of other archival research during the Aamodt adjudication, a document was accidentally discovered (in Mexico) that supported Cochiti's long-standing claims that an area around Santa Cruz spring had been unduly severed from their Pueblo League property. Their claim was later restored by the governor and New Mexico senators.[37]

By the late 1970s, proposing, passing, and getting approval for new dams in the western United States was increasingly difficult. The public's tolerance for cross-

channel megadams had waned. Along with the rise in federal command and control legislation like the Endangered Species Act and the Clean Water Act, civil society veered in another direction to set water aside for rivers and nonhuman nature.

<h2>WATER HANGOVER: THE LEGACY OF THE RIO
GRANDE PROJECT</h2>

Perhaps the most problematic current adjudication for complexity is the ongoing
case on the Lower Rio Grande (LRG; see map 12), which includes New Mexico's
largest federal irrigation project, the Elephant Butte Irrigation District (EBID).
Capped by the dam of the same name, along with the later-added Caballo Reservoir, the EBID provides hundreds of farmers with water on a contractual basis.

For decades prior to the adjudication of the LRG, there was some question as to
whether the federal Bureau of Reclamation ever established its own water rights
claims at the time of project construction.[38] But the larger dilemma involves the
ongoing legal challenge between New Mexico and Texas, with the EBID squeezed
in between. Serving out water to ninety thousand acres, the EBID is stuck between
two state entities that each want more control of Elephant Butte Dam waters. "Tex-
ans thinks Texas begins at the foot of Elephant Butte Dam, for all intents and purposes," a bitter New Mexican farmer told me, referring to Texas' large claims on the
water. I knew I had entered a hornet's nest back in 2010 when the first no I received
for an interview request came from this region. "Too much is at stake," I was told
in an email response, especially for the federal irrigation district (EBID).

Rebecca Snowe, a pecan grower, is part of the EBID and concerned about the
other demands and power players on the river. Some eight thousand people
depend on the water stored in Elephant Butte Dam. Few of them have had a full
allocation of water in years. "The situation is pretty dire, with little water in the
[Elephant Butte] dam year after year—less water for us, less water for El Paso. It's
no wonder the states have been suing each other for years. Plus [sighs], there are
the minnows upstream that get some water before we ever do, along with the birds
in the [Bosque] national wildlife refuge."

As we walked along her groves of pecan trees in the summer of 2011, it was clear
that she did not see quick improvements or solutions. "Our family has been here
for sixty-some years, and I don't want to leave. I don't want to stop growing. Everyone blames us for planting pecans, but these are the best cash crops we have for
down here. What right does anyone have to tell us to stop growing these nuts?" It
should be noted that pecan trees, like almonds, require greater duties of water to
keep them alive. They cannot be "fallowed" like an annual crop. We talked at
length about the similarities of blaming almonds in California and pecans in New
Mexico for perceived water misuse or waste. "And yet people keep wanting more
nuts, right?" she asked pointedly.

Rebecca worried about the future of farming, "yet it is hard moving past what people see as their local heritage food, you know? But the [Elephant Butte Irrigation] district is in a hard place right now, stuck in legal fights, with almost no real storage capacity behind the dam. And then on top of that you have the city of Las Cruces wanting water for its growing numbers. I don't really see a future for a lot of farmers in this valley if we don't get more water down here. Or we may just face more pressure to sell that water to the feds for fish, birds, and the damn [Rio Grande] river compact."[39] As the state continues to sort out the water accounting in the LRG adjudication, the situation has not changed for Rebecca and her fellow farmers. Elephant Butte has rarely been above 10 percent storage capacity over the last decade because of a long-term drought that only loosened its grip somewhat in 2014.

The LRG adjudication suit was requested by the EBID in the mid-1980s yet only became an active concern for the OSE by 1995. Privately, one former attorney for the OSE confided that "our office [OSE] wasn't really ready to staff the LRG adjudication and we knew it was going to be a mess, so we held off on addressing to get staffed up and get proper resources. We knew it was going to be a pretty tough and a long dog-fight. And the recent suits between the state [NM] and Texas haven't helped either, just fighting over dam operations. . . . I know it's going to be a long haul."[40]

Twenty-something years later, the suit has involved five judges, at least twenty state attorneys working for the OSE, and some eighteen thousand individual claimants. Judges in adjudication suits are often not trained in water law. They are typically fine administrative justices, but water is different, and the lack of water law education can be striking. Judges know this, but it is one of the weak points when judgeships rotate during adjudication proceedings. As one former judge who oversaw proceedings of the LRG adjudication for a decade noted in a grand understatement that "the number of judges and attorneys for the state involved in adjudications over a considerable span of time can cause inconsistency" and that because "of this turnover, historical memory can be lacking."[41] New Mexico and Texas continue to argue about historical dates and compacts on the Rio Grande and the Pecos Rivers.[42] The LRG is the current largest "test" of the adjudication process, certainly in terms of acreage and water, in New Mexico so far but may not end up being as expensive as other suits because no claims to date involve the Pueblo or other tribes.

The urbanization of water seems inevitable in the West. Some 80 percent of residents of the western states live in cities. However, in New Mexico, urbanites use only a little more than 8 percent of the state's water. Nearly 80 percent of water still goes to irrigated agriculture.[43] There is no doubt that the "buy and dry" policies of cities have targeted senior water rights in agricultural areas.[44] The data suggest, however, that cities will not have to drain every agricultural patch even in this dry state because any portion of transferred agricultural water goes a long way in the city.[45] Fair, alternative leasing mechanisms for water could help keep farmers on the land yet allow them to occasionally lease water to cities. Perennial water rights may become more flexibly seasonal. The ignored Rio Grande main stem poses significant adjudication challenges. Since adjudication was initially created to count agrarian water, it will be an administrative and political challenge to count the urbanized waters of the MRG.

The ongoing LRG and related Texas-New Mexico struggles over accumulated groundwater use and dam releases offer a preview of what is to come. There are two likely possibilities for what is to come: more settlement agreements and more federally and state-financed water projects. The first will hopefully resolve some of the political tension, while the second will depend on generous funding to quell the unhappiness and compromises generated in the political process. Negotiated settlements have resulted in more hardware infrastructure to meet the needs of these legal compromises. The era of big water infrastructure is not disappearing, but the nature of that infrastructure may shift from surface storage to interconnections between basins, groundwater, and off-channel storage strategies. Like settlement negotiations themselves, this new accommodating infrastructure may be less visible to the public.

The remaining, incomplete work may rewrite not only the allocation of water in the state but the very future of upstream New Mexicans who depend on water. Cities grew quickly over the last century, at the countryside's expense, often based on mythical and nonexistent prior legal bases. Those peri-urban irrigation districts and ditches often paid the price of urbanization with their rural water. Outright urban water dispossession, key disruptive events like new dams, and a major world war played roles in the quiet movement of water to the cities of New Mexico. How the MRG, with its conservancy district and Pueblo populations, will be sorted for water, justice, and rights is a standing mystery. Since these water-accounting procedures have not begun, it is hard to plan for existing city demands and new demands on the rivers. Known and unknown challenges, however, like climate change, endangered species, and new risks to water infrastructure, lie ahead.

9

———

Beyond Adjudication

Nature's Share of Water

*In the end, the guy [state engineer] is going to realize that sharing water is the
only solution we have. He's not going to kick people out of the state because
there's not enough water. He doesn't have that power! He can pretend to be in
charge and know all this stuff, there's so much water here, there's not enough
water there, you know the rest . . . but right now he has to keep issue groundwater permits and that's what will probably force him to change the whole
system. The system has to give at some point, because the weather's only going
to get worse, and it will just keep getting drier and drier, right? So what then?*
—WILLIAM BENAVIDES[1]

Is adjudication and its detailed water accounting for naught if there will be less water
in the future? The dead yet detailed linen maps of past adjudications fix a number to
water that is always moving and seasonally changing. While many water users have
expressed confidence that their local ditches will cope with increased aridity and
rainfall variability, as they long have, they remain troubled by the state's shifting
plans for future water use. The state must create mandated regional water plans,
account for recreation, and create ways to protect federally listed endangered species
like the Rio Grande's silvery minnow. To add further complexity to future water div-
vying, New Mexico, like most of the Southwest, is expected to face water shortages
under the modified climate regime some have termed the Anthropocene.[2]

The Anthropocene refers to our period in which human activity has come to
influence the earth's climate and atmospheric processes. The climate is changing,
and in the Southwest, the predicted changes are a worst-case scenario for water
planners. As William deBuys has eloquently summarized, the western United
States can expect to become drier and warmer, with forests shifting to more arid-
like scrublands. Wildlife will also be affected. A few generalist species will benefit,
while others may decline or disappear.[3]

Water managers a decade ago were already planning for changes in snowpack, spring runoff, and rainfall.[4] Drought gripped the region in 2002 and lin-

gered, drying reservoirs and riverbeds and acequia ditches and prompting senior water rights holders to demand their priority access rights. Study of the past
may give glimpses of the future. Paleoecological studies have suggested that the
western United States has experienced spectacular flooding at times but has
also been much drier in certain periods, including during megadroughts. By
megadrought, climatologists mean not just a dry year or two but dry cycles that
can stretch for decades. Multiple known megadroughts from the relatively
recent geological past serve as an omen for what human-induced climate
changes may force water managers to anticipate. In the third century CE (200s),
geological records on the Colorado River show a roughly sixty-year drought.
Fast forward to the 1100s, and a century later to the mid-1200s, and there are
two more examples of decades-long droughts. These latter two are thought to
have contributed to the collapse and reformulation of the Ancestral Puebloan
cultures in the Four Corners region where Arizona, Colorado, New Mexico,
and Utah now meet.[5]

Imagine the Rocky Mountains, the Sierra Nevadas, and the Cascades mountain
range with 10 to 50 percent less snowpack. For Southwest residents, that is not a
difficult exercise. Now imagine the rivers and streams, including the Colorado and
the Rio Grande, with half their snowpack-derived streamflow. This may become
reality again.[6] As the latest Intergovernmental Panel on Climate Change report
concludes, New Mexico's dependence on the Rio Grande and its tributaries may be
tested severely in the coming decades. By midcentury, the state may have 10 percent less snowpack; by the end of the twenty-first century, it might have 30 percent
less. Worse, these may be conservative estimates of snowpack loss.[7] A recent study
by the Bureau of Reclamation and the city of Santa Fe estimated that San Juan-
Chama Project deliveries may decrease by 25 percent due to these climate factors,
and the city will have to plan accordingly.[8]

The most prudent water managers and planners in the Southwest plan for the
worst—and have done so for the last twenty years.[9] Map 13 illustrates one version
of how the Bureau of Reclamation hypothesized the near future of water scarcity.[10]
The darkest intensity of gray shading connotes the highest possible levels of conflict. The zone of most concern for New Mexico is Albuquerque and the Middle
Rio Grande Conservancy District. Worse, this 2003 map did not account for
already existing climate-related water losses.

While map 13 presents only hypothetical regional scenarios, it does offer cautionary lessons for the Middle Rio Grande, even before climate-change scenarios
are included. Water infrastructure in the twentieth century was designed to regu-
late natural water flows and slow the downstream movement of spring thaw water
and floods. Yet many of the twenty-first-century climate scenarios for the western
United States suggest that water managers may have to alter how they approach
and design water storage. Instead of traditional concrete reservoirs, states and

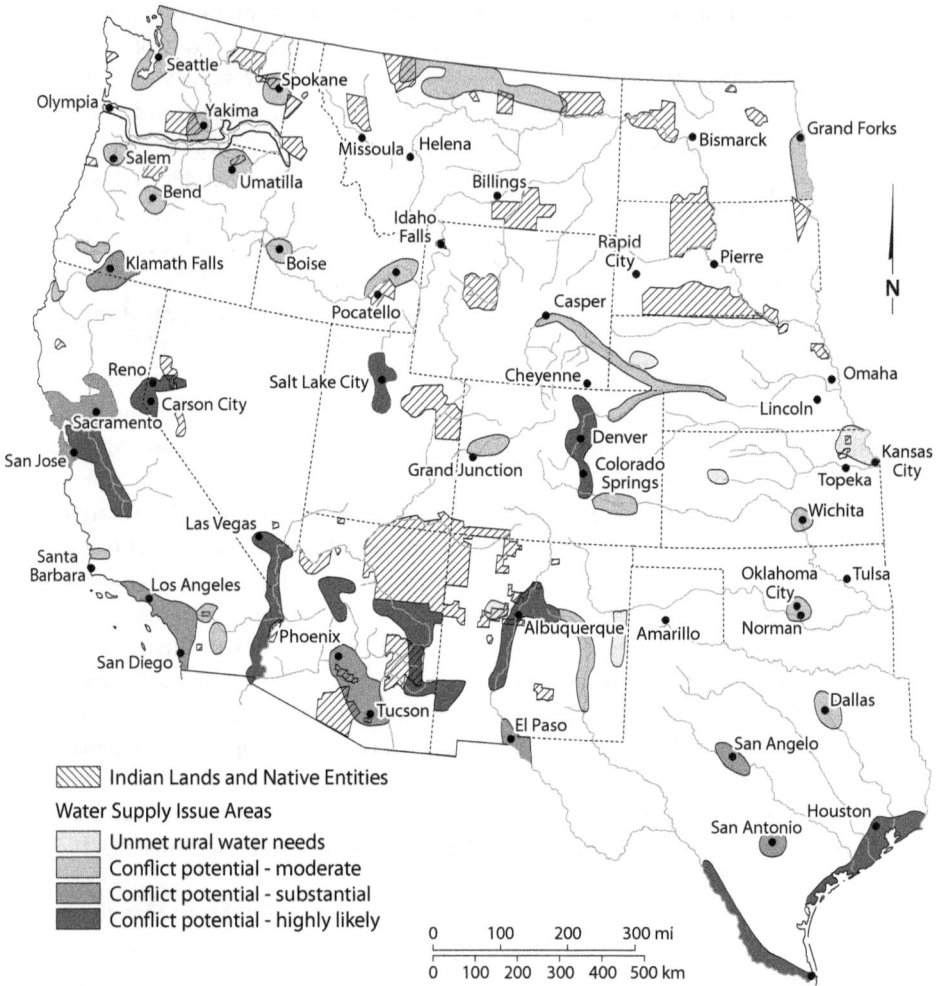

MAP 13. Hypothetical future conflicts in the western United States over water resources. Note that in New Mexico, much of the Rio Grande main stem is in "highly likely" or "substantial" likelihoods of conflict. Adapted from the Bureau of Reclamation (2005).

agencies may have to store water off-river if snowpack is less dependable, and sudden rainstorms become more extreme and unpredictable. Mass storage will then be less about snow and slow melt and more about the clever seasonal storage, above and below ground, of falling rain or sudden flood events. The record of mining aquifers and depleting groundwater may already be changing as cities across the West and Southwest engage in groundwater recharge experiments.

Most water-allocation calculations, arrangements, and even water laws were built on assumptions of water stationarity that long-term water and climate averages hold steady over time. We know this is wishful thinking, and it was never true. Political demands over river compacts demanded an average for old water allocation assumptions. The western United States has always been prone to distinct and variable climates and extreme weather events. This continues to be true. Whether existing water laws and statutes can cope with new water demands—from climate change, endangered species, and rivers, among others—has yet to be determined.[11]

DOES CLIMATE CHANGE "CLAIM" A WATER RIGHT?

In a warming, drier climate, whether on the short scale of a drought or in a prolonged desiccation period, irrigators, mayordomos, and commissioners will face similar challenges. They are already making choices, some of them based on the experience with the 2002 drought. There is much creativity left in ditch institutions across New Mexico and much to learn from their experiences.[12] Ernesto, a mayordomo on the Gallinas River, near the small city of Las Vegas, argued that acequias already knew how to deal with drought: "Look, we have been dealing with drought all of our lives, you know? We know this problem. We live it . . . so we shared the water, whatever we have. If there is none to share, people understood what it means to have a dry ditch."[13] Ernesto shrugged repeatedly in trying to emphasize this acceptance of what it means to live within limits set by rainfall, set by the river. When water experts talk about the sustainability of water, they often mean sustaining current water uses, not the "sustainability" of the resource itself. Ernesto's attitude approaches a true measure of locally sustained and managed water. It's not fatalistic so much as it is realistic.[14]

To be sure, this scale of ditch pragmatism is difficult to upscale and translate for a city, even a small one like Las Vegas near Ernesto's acequias. Las Vegas is not large, but its population grew while the water supplies did not. Like many cities in the arid Southwest, Las Vegas seeks out additional water supplies while exploring tougher institutional choices: conserving water, pricing water more aggressively, or rationing water. The urban politics of augmentation, of always seeking out "new" wet water, seems self-defeating. Adding water encourages growth and has served as "fuel for growth," as one author has convincingly argued.[15] Climate change, the Anthropocene, and drought all provide productive excuses for conversations about the future of water. These anticipated decreases in water ability provide opportunities for dialogue across sets of expertise.[16] Nature will always be the final judge of water adjudications and sets the hard limits of water from year to year, even if groundwater pumping has long been used as a temporary cheat of this final judge.

Fire, ash, and sediment will also be new concerns. The 2011 Las Conchas Fire was a reminder of how climate and fire may extract new water demands and have an impact on water infrastructure in the long term. By its end, this devastating wildfire had charred nearly 160,000 acres of upland conifers, piñon-juniper forest, and dry grassland. Severe fires like Las Conchas remove most of the standing biomass and render soils nearly impermeable to rains that follow. Without vegetation and with little water infiltrating the soil, even a small rainfall can be converted to a flash runoff, thick with ash and soil as floods roil downslope. In the case of the Las Conchas burn scar, the burn and ensuing runoff also exposed a toxic World War II legacy: nuclear waste unceremoniously dumped by the national labs at Los Alamos.

In recent years, dozens of wildfires have erupted across the Southwest, many at wildland–urban interfaces, intensifying the urgency and costs of firefighting efforts.[17] Westerners now have an unwelcome fifth season from April, May, and June: the fire season. Fire might seem tangential to water adjudication. However, firefighting requires water, and forest fires have an impact on sediment loads; infrastructure; and ultimately, water supplies and water quality.

When I revisited the Buckman Direct Diversion Project in 2014 with a group of students, the manager affirmed how difficult managing sediment and sand had become since the 2011 Las Conchas Fire. As we approached the holding ponds for raw water upslope from the Buckman Dam, at the water treatment plant itself, the project manager shook his head grimly. "Well, that's *no bueno*," he said, as we stared at concrete tanks that appeared to be holding chocolate milkshakes: muddy sediment rimmed the side walls twenty feet high. The dam design at Buckman was supposed to be less problematic for managing the bed load from the river. However, the new intake dam is also at the mercy of the river, which delivers more than just water.

Water managers at Buckman have additional worries in treating drinking water supplies for the city of Santa Fe. The diversion project is located across from the canyons that drain out of Los Alamos National Laboratories, where irradiated waste was dumped for decades. This is on top of the naturally high plutonium levels in the Rio Grande.[18] Just like the hundreds of ditches that require a spring clean-out *(la limpia),* large-scale water infrastructure must also cope with the material reality of moving water *and* its associated bed load. Fire, flood, and sediment have their own reciprocity. Autonomous and variable water was not accounted for in the adjudications, only fixed notions of human water ownership and land use. The land cover, the vegetation, and the climate variability of any basin were all ignored.

Including a water rights mechanism that anticipates climate change–related losses of water would have to work in two ways: First, it would have to account for stream losses due to aridity, a water right *because* of climate change. Second, it would have to set aside water in particularly vulnerable settings, a water right to *combat* climate-change effects. The first would account for snowpack declines and increased evaporation from large reservoirs. The second would acknowledge that water is needed, for example, to put out fires in critical environments because of current and predicted rises in future temperatures. Most would agree that saving a house or town is a beneficial use of water, although this is not captured in water codes in the western states. These new problems get to the root of whether old codes, focused on largely agrarian "beneficial uses" of water, can be redefined for new challenges.

The OSE has taken various measures in the past twenty years to plan for shifting water supplies and water priorities. One of the novel state tools is the *strategic water reserve*. The origins of the reserve date to the Pecos River settlement, which included a struggle between New Mexico and Texas and involved the Carlsbad Irrigation District.[19] The purpose of the reserve is twofold: to acquire water rights for the silvery minnow (under the US Endangered Species Act of 1973) and to meet the interstate compact obligations with Texas downstream. An optimist might view this as a good governance tool reflecting sensible ecological economics or trying to benefit ecology and political economy simultaneously. A cynic might call it "another tool to avoid getting sued," as one attorney put it to me.[20] The truth lies somewhere in between.

The problem that New Mexico faces is that the strategic reserve program is severely underfunded. It may also be underwatered, with not enough actual water allocated to make much difference in any given year. The New Mexico state water code still makes no formal and explicit recognition for "environmental flows," even though they are mandated by federal laws for species protection.[21] Despite its deficiencies, the reserve has become the de facto pool of water for Endangered Species Act purposes and could be creatively used to meet these additional climate challenges in the future.

NATURE'S CLAIM TO PRIOR AND PARAMOUNT

Among the most recent pressures has been the push to extend water rights to nonhuman nature through both the US Endangered Species Act and from local environmental nonprofits that continue to sue to restore water to particular rivers. This has happened under a variety of guises. In many cases it means leaving water in a river for the river itself or to preserve threatened species or environments. However, the notion of extending rights to nature in the form of an ecological governance model for active water management is not without controversy. When New Mexicans have been historically excluded from their *own* forestland and waters, they are unlikely to feel sympathetic to trees, fish, or nature in general.[22] While ecological and ethical water governance is laudable, and in many areas sorely needed, there are historical, political, and legal challenges to implementing a rights-based approach for nonhuman nature, especially in a poor region.[23]

Some other western states, like Colorado, have had instream flow programs for decades. However, New Mexico has no such formalized program and only grudgingly "gives up" some water for nature, for the river, or for endangered species. "When the feds are watching," a local river activist observed back in 2012, "then the state feels some pressure to respond."[24] In the late 1990s, under pressure from the US Environmental Protection Agency and the US Fish and Wildlife Service, Attorney General Udall offered a legal opinion regarding the silvery minnow along the Rio Grande near Albuquerque and points south of the city. It was the first time that any New Mexican official had recognized putting water back in the river for an endangered species as a kind of beneficial use of water.[25]

The first decade of efforts to preserve the endangered Rio Grande silvery minnow was marked by some failed experiments, including the rather tragic Pete's Pond silvery minnow habitat microfarm. Built with fanfare in the early 2000s, the test pilot project established a set of pumps and pools on the eastside floodplain of the Rio Grande, just south of Albuquerque. The idea was to create a minihatchery from which silvery minnows could be released. Millions of dollars were spent. However, when the water was finally circulated into the floodplain experimental area, the flow quickly disappeared; it seemed that no one had bothered to account for the sandy soils. The original conception for the project folded quickly.

But this early failure at ecological experimentation and restoration provided important lessons. Learning from that initial mistake, a new location was chosen in the small town of Los Lunas, where a current silvery minnow *refugium* now raises the fish for release along the Middle Rio Grande. Pete's Pond may also be resurrected in some new form, focusing on other species needing help on the main stem of the Rio Grande.[26] Ecological water can produce new downstream uses, help the states meet their targets under the Endangered Species Act mandates and goals, and start to shift notions of what the beneficial use of water entails. It is not just about who gets water rights, but what.

Twenty years later, the state's formal policy language remains simply a former attorney general's opinion. Yet the state created funding and institutional mechanisms to allocate, purchase, and move water for species and wildlife programs despite the lack of a state water code mandate. Recently, New Mexico ranked third (behind Colorado and California) among the Colorado River Basin states on transfers of water for environmental purposes. Implicitly, the state has adapted its own code to the new demands of endangered species, but state statutes remain silent about the rights of other species to water.[27]

Some of the return water that sustains endangered species, ironically, is quite human water. Albuquerque's wastewater treatment plant sends some 100–160 cubic feet per second (cfs) of water back into the Rio Grande, some 200 to 315 acre-feet per year (AFY) per day moving downstream. Water pundits like to joke that this wastewater is the fifth largest tributary of the Rio Grande. However, as a plant operator pointed out to me in 2011, "It [the wastewater] may smell funny, but there's nothing funny about how it helps Elephant Butte Dam or those silvery minnows."[28] Return human flows out of the Albuquerque wastewater treatment plant help farmers and wildlife, as well as the state's efforts to meet Rio Grande Compact obligations to Texas. Timing those flows and treating wastewater for reuse may become more critical to New Mexico (and Texas) as water shortages become more common.

One of the repeated clichés in environmental history and natural resource law is that "nature bats last." The sports analogy is crude but another reminder that nature is the ultimate arbiter of our human conventions. It is also true in another way: nature often gets its share of water last. Human uses still get first dibs. The creators of prior appropriation were not thinking about fish. They were concerned with human claims to water for mining or agriculture. Changing the saying to "nature bats first" in the twenty-first century may be a better starting point in planning for a future with much less water. A more realistic analogy would be that nature bats first in terms of what is available in snowpack and rainfall, bats mid-lineup in summer during monsoon rainfalls for agricultural and urban purposes, yet only occasionally gets a swing at the end, when it comes to any water left for rivers and particular endangered species. New Mexicans have to manage the water available, not the water hoped for in best-case scenarios. Most water managers are well aware of this. That remains true even as constituents constrain how creative managers can get.[29]

New Mexico recently emerged from a ten-year drought that was difficult for water users, managers, species, and waterways alike. In 2002, whole streambeds went dry on rivers that are typically perennial. Farm fields were fallowed, dairy cows were sold, and hay had to be imported into the state. The drought, as discussed in earlier chapters, prompted New Mexico to implement new forms of water management before the long work of adjudication was finished. It was a stark reminder about the ultimate availability of water in the greater Southwest.

As I showed in previous chapters, rivers do not only run dry because of drought. As Santa Fe disconnected the natural flow of its namesake river, the river was put through pipes and taps. A dry riverbed ran through the heart of the city through much of the twentieth century (see figure 11). Human demands deplete waterways, but they can also make them flow again. The golden era of environmentalism (the 1970s) produced pressure for states to put water back into the rivers. Watershed politics in the twentieth century have reshaped the Santa Fe River in new and surprising ways. Santa Fe has used its municipal power to exercise a different kind of sovereign choice. It has put water back in the Santa Fe River.

This was done through citizen pressure and a dedicated watershed association group. The "living river" program was devised to return some flow to the river's native bed. The allocated 1,000 AFY is minor in terms of volume but symbolically

FIGURE 11. The dry bed of the Santa Fe River, in 2009, prior to the Living River program enacted by the city to release up to one thousand acre-feet per year of water to the channel. Photo by Ann Perramond. Used with permission.

rich, reflecting a new appreciation for what the river means.[30] With this program, a trickle of water again flows through the watershed at key times—not enough to reach the Rio Grande but sufficient to make the river seem alive again, at least for Santa Feans and tourist eyes in the heart of the city.

What is perhaps astounding is that the program was implemented during the drought. In the dry summer of 2014, along with students, I toured the city's main reservoirs, which stood empty because of needed repairs to a water tower. A Santa Fe city water division employee commented on the living river experiment, with the dry reservoir as a backdrop. "It's interesting, sure," he said. "But if Santa Feans really want water in the river, it's simple: we can just drain the reservoirs upstream. Then what? We won't have any water in the summer, that's for sure. This program won't operate unless we have good water years. Tourism is too big an industry to dry out the hotel room taps."[31]

Moments of water crisis like the 2002 drought at least provoke a new kind of thinking that can shift governance for more flexible water use.[32] Santa Fe's living river shows what cities with strong water utilities governance and citizen leverage

FIGURE 12. Released water in the Santa Fe River channel, summer 2015. Photo by the author.

can do once they understand their water supply options.[33] Such experiments remain limited in their scope but critical for imagining new water policies that address human and nonhuman needs, including other species and rivers.[34] Santa Fe dried out agrarian ditches in the late nineteenth and early twentieth centuries. It progressively moved the Santa Fe River into pipes and taps, ignoring citizen and irrigator concerns. In the new century, it reversed course under citizen pressure to put water back into the river for nonagrarian, ecological purposes (see figure 12). Political pressure in progressive places is not the only option.

Water trusts are another increasingly common way to shift the water balance among farmers, species, and rivers. Typically, with water trusts, nonprofits purchase and then retire water rights from human uses so that the water can return to rivers. Water trusts are operating in various basins across the western states. In the

Southwest, specifically, a major effort is underway to acquire ecological water rights through a nonprofit partnership (Colorado Water Trust) working in tandem with the state of Colorado's efforts to acquire instream flows. More ambitiously, a group named Raise the River is attempting to rewater parts of the Colorado River Delta, long a dead zone sacrificed to human withdrawals.[35] While the command and control approach under the Endangered Species Act is still active in western waters, these market or economic incentive–based instruments are also putting water back into rivers on a voluntary basis. Whether market-based or nonprofit voluntary "donations" of water can be sustained is questionable. There are few well-funded state programs that put water back into streams. These measures, nevertheless, represent a shift in the use of private water rights toward a larger public trust purpose.[36]

These demands for ecological water play out in a state with a long history of human water use. Even a century ago, New Mexicans worried about whether there was enough water to go around. Today such concerns are heightened, even without accounting for the changing demands of nature and species. At the January 2018 New Mexico dialogue, I listened as an earnest US Fish and Wildlife Service officer summarized his small group discussion, starting with, "We all agreed at our topic table that water for ecology and other species was important . . ." At a nearby table that had been discussing water rights for farmers, an older gentleman wearing a cowboy hat overheard. He scoffed and chuckled, shaking his head. Clearly, New Mexicans disagree on these new directions, with many expressing agrarian skepticism of ecological purposes, or "water belongs with the land, period," as one irrigator put it stridently.[37]

ADJUDICATION AND INFORMING WATER FUTURES

In urban areas, it makes sense to price water (and water use) rates in an aggressive, progressive fashion. Urban water users are already used to paying for water access. What urbanites are actually paying for is the delivery system. The water itself is dirt cheap. Markets do have the ability to change individual behaviors, even if mechanisms for pricing are never perfected. Rural residents, however, fear the capital heft that cities carry in offering lease rates to farmers, tempting them to send their water away from the land. Similar concerns stem from programs to "retire" agricultural water and put it back into rivers.

For the last hundred years, the state zealously tracked water users to try and account for any water left over. New demands to leave water in streams for nonhumans are unaccounted for in the adjudication process. Climate shifts and water scarcity are also "shares" of water that remain a blank data set in the state's adjudication process. With these new demands, the purpose of adjudication may be subtly shifting. Instead of hoping to allocate more water than may have ever existed, the state could view adjudications and their data sets as tools for planning a fully allocated state.

In New Mexico, as in countless other semiarid regions, the relevance of water supply, provision, and conservation has only increased in recent decades. Far from Colorado's snowpack, too close to Texas' demands for water on the Rio Grande, the state faces limited options for magically creating water. All water use, like politics itself, remains local.[38] Water remains embedded in its cultural contexts and places. However, the new regional, translocal associations and pressures do not necessarily translate to participatory water governance or planning on the regional level, as discussed in earlier chapters. The state might consider how all of the data collected during adjudication could help its own citizens' plan for their own water futures.

If, for example, the adjudication data were live linked and more transparent, state agencies and engineers could use this information to model future climate and water-demand scenarios. They could also better plan for worst-case scenarios and more flexible water leasing, as Leandro on the Embudo River suggested. "We kind of know already there's no more water here to share or give to new users, so why not just start to plan for the worst to come with this state stuff [adjudication]?" he said. "All the demands for fish, and ecology and things—there are better ways to do this. Look at how cattle ranchers get paid if a wolf takes one of their cows. We could benefit from leaving water in occasionally . . . if someone paid me, I'd think about it. Seriously."[39] To date, the data have primarily flowed uphill to the state's agencies, with little coming back down for useful local implementation.

Tonio brought this up in the summer of 2011, as we walked his ditch along the Mimbres River. "I mean a lot of us have smartphones. There has to be a better way to get some information in our hands, too. They [OSE] can't just meter us, slap meters on our houses, and then turn around and refuse to share information. It's just stupid, that water information, even the [water] rights information, that should all be much easier to use, easier to access too . . . that's part of the problem. The state just takes and takes information but never gives us back . . . anything . . . about what they're doing."[40] This issue was also palpable during the 2017 New Mexico Water Dialogue, during which participants voiced concern about the lack of information available for citizens from the OSE.

Many New Mexicans I spoke with over the years also wished that the state engineer would focus on gross quantities of water, ditch and stream inputs and outputs, instead of expending so much administrative and court time quibbling over individual water users and subjecting them to years of litigation. As Alejo from near Las Vegas stated, "I mean these guys [OSE employees] are running around trying to count all us small-timers. Meanwhile, we know all that water is evaporating, going to Texas, getting used in new domestic wells . . . all those suburbs outside Albuquerque and Rio Rancho? You think they are not using more water than my small acre here in the middle of nowhere? They're wasting their time and ours. We were here before the state was; they should be counting all the big water that goes to the cities and on the rivers. All we do is send that water, anyway, downhill towards them."[41] Focusing less on water users, and more on water volumes and movement, could be more productive to solving problems. This is especially true for the OSE, dam operators at the Bureau of Reclamation and the Army Corps of Engineers, and water managers in irrigation districts, given climate projections and the implications of paleoclimate data.[42]

Because the agency of water as a "property" tied to individual bodies ("owners") creates problems for the adjudicator and the adjudicated, moving away from tracking individuals to tracking water as the object itself might be a less adversarial way of dealing with water in a warmer, drier century.[43] Most "expert" water management is done with meters, switches, gates, and buttons. Cities are already using improved metering technology to create a more closed-loop water system in order to avoid leaks.[44] Keeping better track of the aggregate flows, too, might help the state avoid the worst of western water practices. Maybe there's a larger purpose and use for adjudication in the reams of paper, the terabytes of information, kept by the state's agency. The OSE could start to estimate whether or not the state has overallocated their waters and by how much. Instead of just asking for regional water plans from New Mexicans, as Alexander, a key participant in one of the regional water plans in northern New Mexico, exclaimed, "Give us information to help us plan for our own water future!"[45] Making these data more public, as dozens of interviewed New Mexicans argued for, would also allow other states to gain comparative perspective for their own challenges.

The state could reframe the adjudication "product" as a way to better account for, govern, and plan for future water use and water scarcity. Whether it can move away from adversarial court procedures or enforcing the priority administration of water rights is a bigger challenge. It will take a completely new culture of water to reimagine how the state can support farming, cities, and industry with less overall water while serving unknown future needs during this new climate era.

COMPARATIVE WORST-CASE SCENARIOS

Hector, whom I walked with in the introduction, laughed about all of these new competing demands for water as we strolled along his ditch years later on a return visit. As a mayordomo, he had already felt the wrath of his fellow irrigators during the bone-dry year of 2002 and was not looking forward to more years like that one. But he also knew that dry years were becoming more common. "Scarcity is the law, not that code [state water code], and if there's no water, what is the point of water law, you know? What are they gonna do, adjudicate *zero* water? Are they gonna show us our dry ditches and say 'cooperate over that?'"

Adjudication as a process, as Hector suggests, may be penny-wise by tracking individual water rights but pound foolish when it comes to aggregate water quantities, much less the realities of twenty-first-century water scarcity. Hector followed with a great analogy: "It's [adjudication] like counting all your pennies in a piggy bank that moves, disappears, or changes hands every year. It's always different, never the same." Water continues to move beyond any state's view, purview, or sovereign control. It is a fugitive resource that mocks any attempt to make it "stationary" in the sense of water rights stuck on a dead linen map. The water moves, the data do not.[46] The numbers given through the duty of water may not have changed, but actual water certainly has.

If New Mexicans want to know what the future could hold, all they have to do is look at California. California has almost four times as many paper rights to water as there is actual wet water.[47] In July 2014, the state of California warned that thousands of water rights holders with priority dates after 1914 were to be cut off due to the historic drought. Fields and canals went dry. Orchards were chopped down.[48] The recent 2016–2017 deluge in California has given the Golden State a reprieve on making water scarcity decisions, although it is short-term relief. Flooding and devastating wildfires have followed, affecting water quality and infrastructure. The potential lessons worth learning here are twofold. While California has massive political and economic clout and senior water rights claims to an impressive share of the Colorado River, New Mexico does have at least a systematic way to account for its own water. California does not.

The California dream of a large-scale, long-term solution has not stopped smaller experiments, either. In August 2015, for example, a local reservoir in Southern California was filled with black plastic balls in order to lower evaporation water loss in an urban water district and to "keep the birds out of the reservoir."[49] New Mexicans, like most westerners, also keep hoping that technology will fix "it," when the "it" is us—our limited set of imaginations and political willpower when it comes to the vital liquid that all living things need. Planning around worst-case scenarios is one place to start. Technology can help managers and planners model, or formulate, scenarios for less water or at least fully allocated waters.[50] But technology cannot make decisions without the consent of New Mexicans.

Paradoxically, in California water management has been so unmanaged that a process as complicated and byzantine as adjudication is now seen as a potential solution.[51] Let me say that again, because it is almost unfathomable: things were so bad in California that they thought *adjudication could help simplify things*. Such a move seems a folly, at best. As legal scholar Lawrence MacDonnell recently wrote, "There is something audacious and, at the same time, almost foolhardy about undertaking a general stream adjudication at this stage of water development in the West."[52] Adjudications simplified nothing about water in New Mexico, and Californians would do well to ponder that. Both states will be challenged to sort through how climate shifts will reduce water supplies, increase fire frequency, and create water infrastructure challenges from sediment in the long run.

Nature continues to claim its own share of each state's water supply even if adjudication was not designed to count nature's share. Planning for worst-case scenarios, higher evaporation, and less water and snowpack is a good starting point for all western states.

Back in Las Vegas, New Mexico, William Benavides and I leaned on a locked headgate one November afternoon talking about the implications of farmers fighting cities in dry years. He thought the inflexible, top-down approach that the state used to count water was no longer useful. "Instead of having us jump around proving our water rights and adapt to adjudication, maybe it's time for adjudication to adapt to the new climate realities we all face."[53] The challenge for New Mexico and most western states is whether nineteenth-century prior appropriation water law, optimistic twentieth-century water codes and infrastructure, and twenty-first-century climate takings can be reconciled.

10
———

Water Coda, with No End in Sight

Understanding the landscape means more than looking at a map.
—LUCY MOORE[1]

Water, in rivers and connected aquifers, retains its own wild autonomy outside the legal system, outside of state-mapping enterprises and accounting procedures.[2] Water continues to evade, move, and leave state boundaries. This mismatched understanding of how water use over time and space is changing but perhaps not as quickly as desirable.[3] Adjudication maps cannot capture how people feel about water and place. Adjudication attempted to simplify the places, spaces, and quantities of water in a two-dimensional way. What adjudication could not do was simplify the dimension of time—the long-accrued historical experiences and meanings given to water by the different water cultures and sovereigns in New Mexico. The state compressed space in maps but could not compress time and accelerate adjudication. Changes in water use, over time, are also lost in the process.[4] Janet Fresquez, a resident of Los Ranchos de Taos, put this in clearer terms: "They want our water on their map, but they don't care in the end how we think about our water or share it."[5] The state's view of water as a resource or as a commodity is at the heart of why people feel anxious about adjudication. Rural New Mexicans remain fearful that water will leave their valleys and villages high and dry.

Adjudication continues, and there are no easy conclusions. The process went relatively smoothly in areas with few people, little water, and no competing local water sovereigns. In those cases, like the Canadian and Dry Cimarron Rivers, or along parts of the Mimbres, the state could force its redefinition of water use rights as private property. However, in other areas, competition for senior rights and earliest dates sparked tensions between communities and neighbors. In culturally diverse, legally plural valleys, such conflicts were amplified. The process created or at least renewed adversarial water relationships. The state encountered a long-

177

delayed water justice problem when it quantified Native water rights for the Pueblo, Apache, and Navajo Nation. Settlement has been the expensive result of stalled or failed adjudications in cases where long-term social relationships tied to water had to be renegotiated. Hispano and later Anglo-American colonial settlers homesteaded water in valleys occupied by the Pueblo. After both waves of "water colonialism" had taken effect, New Mexicans discovered an unsettled past hard to reconcile.[6] The state instrument of adjudication was poorly designed to address federally defined indigenous waters, much less to reconcile the remaining differences between postcolonial water sovereigns.

The latest water settlements were crafted to escape the state courts. Settlements never satisfy all parties entirely, but they can offer a locally negotiated and hopefully more suitable set of trade-offs in the end. New Mexicans renegotiated the state's vision of water as a property right to one that was more customary, flexible, and less adversarial. Agreements, in the form of negotiated settlements like Aamodt and Abeyta, are no less complicated. They are often more expensive, less "clear" in terms of what they resolve, and may rely on water that does not exist. Federal, state, and local agents must then struggle to match the agreed-upon water arrangements when little water exists to satisfy the compromise. New Mexicans will continue to face the unintended consequences of these long-term agreements that took decades to resolve.

UNSETTLED WATERS: STATES, WATER RIGHTS, AND PROPERTY

Studying how any state approaches and executes water adjudications provides interesting conceptual implications and conclusions. Sociologist Donald McKenzie once argued that financial modeling tools changed the very market behaviors they purported to study and measure. The same can be said for water adjudication: the process of adjudication was the *engine* for creating water rights as property out of public waters. It changed water relationships. It was never the simple *camera* the state thought it was to understand water.[7] In some ways, this should not come as a shock to anyone.

New and imposed law, after all, has often been used to administer a new form of political economy and property rights regime.[8] The twentieth century witnessed a remarkable human transformation of rivers.[9] The large rivers of the West were reforged as organic machines as natural streams were made more mechanical and slow flowing through infrastructure.[10] Less attention has been paid to how implementing a private water rights system changed water use and water relationships between people.[11] When new water law was imposed in 1907, it was simply the latest attempt in redefining water yet again. The new water code replicated a process long familiar to New Mexicans: a "story of colliding property regimes, government collusion, expanding markets, and people's loss," as Maria Montoya argued in the case of land grants.[12]

By its very design, state redefinitions of water as both a publicly owned good and simultaneously a private-use right created the fragmented set of jurisdictions visible today across western states. This conception of water as both public-yet-private property is visible on acequias: the commissioners and mayordomos control access to the ditch rights. Individuals hold private-use rights. The state owns the water. Fragmentation is not just built into competing institutions; it is built into the very fabric of how we now treat water and water rights. Private property-use rights, domestic mutual associations, ditches, cities, counties, Native sovereign nations, individual states that assert states' rights over water, and federal agencies all overlap and create this complicated mosaic.

New Mexico has not completely sold its citizens on the merits of treating private water rights like private land, as a one-dimensional private-use good, because of prior understandings of water. It is also because water is a fugitive substance, essential, and hard to treat as a form of resource "rent" from which value can be crudely extracted. Native sovereign powers like the Pueblo and the constrained local control of acequias, both of which preexisted the state's redefinition of water, continue to challenge this one-dimensional water-resource reading of the state. Laws and private property, as illustrated by the examples in this book, are enmeshed in relationships and slowly deployed, not instantly obeyed or recognized. Most New Mexicans ignored the state, adjudication, and prior appropriation water law until the process was upon them.

The long adjudication process coproduced expert knowledge communities and a balancing of local expert knowledge. New Mexicans bent the arc of legal proceedings, and their collective responses to state adjudication created new conversations and nonprofits centered on water. Local principles of water sharing have been refracted into the Office of the State Engineer (OSE), which at least implicitly espouses water sharing, as well as transparency and equitable treatment.[13] Adjudication has produced more information about water and water users, and that information should be publicly shared. In fact, since the public pays for the OSE's work, it *must* be shared. To that end, adjudication as a kind of forensic water accounting may be worthwhile, just not for the reasons that adjudication was created for originally.[14]

A century ago, every local water sovereign from the acequias, to the Pueblo, to new conservancy districts thought they were in charge of water. That was true, to a certain point, but only because state water law did not take sudden effect on water management and allocation. The state's attempt to flatten out differences by only emphasizing an economic purpose to water resulted in complex and tedious adjudications that reflect those local differences. The choices made will reveal how pragmatic New Mexico and its policy makers can be.[15] Adjudications and water settlements represent distinctive forms of legal-political technology.[16] They were products of, and produced new, infrastructure in the form of pipes, dams, and interlinked regional water systems. As described in chapters 2 through 4, replumbing basins to satisfy settlements is an expensive and complicated way of making legal agreements work in the long term to satisfy the various water sovereigns and cultures in this state. Rivers and basins continue to be adapted and reengineered to satisfy a shifting set of parallel legal traditions.[17] New Mexico's experience with redefining water illustrates that water sovereigns exist in parallel and multilayered universes that continue to complicate two-dimensional state views of water.[18]

A PROCESS THAT OUTLIVES US

"Don't worry, Eric, adjudication will outlast us all," a former OSE employee told me back in 2009.[19] Certainly, the *effects* of adjudication will outlast us. Adjudication remains a complex, expensive, semi-invasive bureaucratic procedure that drafts and recruits all manner of local, state, and professionalized expertise. Settlements do much of the same, without seeming as adversarial, but do not solve perceived problems of representation in the final negotiations. To skeptics and cynics, both processes seem to serve *only* the legions of experts waiting to make a living from the process.

Meanwhile, adjudication-industrial complexes are quietly at work throughout western courtrooms, private law offices, and state administrative agencies. Sylvia Rodriguez had prescient remarks on this industry over twenty years ago, focused squarely on one of its problems: a conflict of interest in making the process more efficient.[20] Why would experts speed up or facilitate adjudications when so many of these professionals derive their livelihoods from the slow, incremental process? Even the optimists among the experts think that it may still take at least fifty more years to fully adjudicate the waters of this sparsely populated arid state.[21] The old analogy about water in the West being structured around "iron triangles" representing agricultural interests, western congressmen, and water agencies can easily be reimagined as a pyramid scheme. One can perhaps appreciate, then, the oft-quoted local irrigators who share some variation of this sentence: "Only the lawyers win when it comes to adjudication."[22]

As the state tries to take a stratigraphic core of water-as-resource in New Mexico through its adjudication process, respective water sovereigns defend their layer, scale, and worldviews of water. Most of the New Mexicans I spoke to felt misunderstood and worried that they had been made enemies of each other and of the state through the adversarial aspects of adjudication. Listening to living water users reveals what is often invisible to official state perspectives and the consequences of the process.

As I've argued, adjudication may not be entirely futile in New Mexico, even though the ultimate purpose of the process may change. There is real value in how New Mexico calculates the consumptive use of water, as discussed in chapter 5. The adjudication lessons from New Mexico can inform other western states, like California or Texas, hoping to account for their own water scarcity problems.[23] Where this more precise approach to water use might pose challenges is if actual consumed water data are *too accurate*. In that case—if the waters are all consumed—future state engineers may face hardened water quantity constraints and consequences in trying to allocate water-use rights to water that does not exist.

In many cases, adjudication seemed to carry no benefit for New Mexicans, other than confirming what they already knew about water use and ownership. The state engineer was mandated by the water code to "know and see" these waters and to firm up their water rights by individual. Settlement, however, comes with real financial benefits in the form of payment, infrastructure, and accommodating language. As I overheard a state official proclaim at a recent water dialogue symposium in January 2018, "These federal resources for water settlements are a vital financial tool for New Mexico, and a real benefit to our residents."[24] Now that residents draw on expertise and the availability of federal resources and funding, it is unlikely that the adjudication-industrial complex can be disentangled from further water settlements. The danger is that water agreements may be predicated on water shared too many ways, or that will be diminished in the future. The future of water sorting may be less adversarial through settlement, but it will certainly be more expensive for both the state and the federal government.

INTERSTATE AND SOVEREIGN VIEWS: WATER QUANTITY AND WATER QUALITY

New Mexico is paltry in its snowpack compared to Colorado and not powerful or wealthy enough to fight off lawsuits from Texas. While employees of New Mexico's OSE have done their best to protect the state's sovereign waters, it is often an uphill battle against Texas. The ongoing struggle between New Mexico and Texas, in the US Supreme Court in 2018, over dwindling water in Elephant Butte Dam highlights that water is not just a single state's affair. As a state, New Mexico is the middle child on the Rio Grande, subject to headwater actions upstream in Colorado and always worried about what Texas has up its legal sleeve. The Rio Grande has less than a tenth of the flow of the Colorado River, an additional factor for New Mexico since it relies so heavily on the main stem of the Rio Grande. This is all challenging without considering that the Rio Grande is a natural border between the United States and Mexico, and occasionally, the rights of Mexico are discussed overtly and tremulously in New Mexico water circles. The paltry shares of water afforded to Mexico on the Colorado and the Rio Grande reinforce political tensions over water. Like the irrigators on a ditch having to share available waters, compacts with neighboring states will have to fully embrace the proportional sharing logic so common to acequias.[25]

Water has no border, in spite of state laws and wildly different state water codes. Water moves. It flows above and below ground. It evaporates and takes other forms. The 2015 release of water down the Colorado, an experimental agreement between the United States and Mexico, resulted in the first trickle of Colorado River water meeting its own Colorado River Delta in decades. It is unclear whether the dam release will be repeated, but it does show what new cultural attitudes toward water sharing can produce.[26] Times of water crisis often force us to connect these dots. As a New Mexican senator said, "Drought serves a useful purpose sometimes, it makes us get together and talk about it."[27] Scaling agencies' flexibility and jurisdiction to cope with longer-term climate adjustments and reductions in snowpack and streamflows, however, will be a more difficult challenge.[28] Transboundary water problems exist not only between political states but also between surface and groundwater institutions.[29]

In *Unsettled Waters* I have focused on the visible and quantified waters of New Mexico and the current thrust of water litigation. Groundwater remains mostly invisible even if direct well metering and remote sensing offer some degree of clarity. As we learn more about the invisible waters, the conjunctive management of groundwater and surface water will be crucial for maintaining base flows in rivers and for supplying cities. New Mexico has a conjunctive water policy for surface and groundwater. However, water adjudication in New Mexico and throughout the West does not address another important aspect: water quality. If people cannot safely drink water or use it on agricultural crops, that water is no longer a "resource," or it becomes a more expensive resource that has to be treated first for human use. Water obeys the laws of thermodynamics; it is matter that cannot be created or destroyed. However, it can become contaminated to the point that it becomes unusable to us and to other species. The next frontier for water managers, politicians, and users is the water quality dilemma.

Water quality has long been a concern for the Pueblo. Over twenty years ago, the Pueblo of Isleta, along the Rio Grande, created stringent water quality rules as a Native sovereign downstream of Albuquerque. Since the city of Albuquerque's wastewater outlet flow was (and remains) the fifth-largest tributary on the Rio Grande, Isleta eyed that water quality carefully. As one past resident of Isleta told me, "They woke up in the city when we created those new standards, even if that court case (Browner v. Albuquerque) created some challenges for us in the long run."[30] The Isleta and Albuquerque case is a cautionary tale for thinking through how cities and farmers alike will have to pay attention to water quality (a federal power) as much as water quantities in the future. Most Pueblo groups are waiting for some degree of water justice in terms of both water quantity *and* quality.[31] How water is produced, used, treated, released, and stored in New Mexico provides lessons for the challenges facing other western states.[32]

SEEING WATER LIKE A MAYORDOMO

In the end, it makes sense to return to the landscape level of water. On my last visit to Hector's small plot on the Rio Embudo in 2014, it was spring and recent floods had taken one of his acequia's diversions. Crumbling concrete and a bent headgate were evidence of the torrential downpour. Hector's thirty-something son, Andrew, shook his head behind Hector's back. After walking back to the family house, Andrew was still looking morose, remarking on the seeming futility of doing yet another repair on that section of the canal since it lay in the pathway of a giant arroyo. The arroyo lies dormant most of the year, but when heavy rains fall, it rages with a flow no one knows how to control. He pointed to the other bank of the river. "We should abandon that ditch and build on the other side of the river."

Hector turned with a distant look on his face. "We tried that, Andrew. Didn't work either. Too steep on the other side, same issues, just more sediment." Thinking like a mayordomo on an acequia has its limits. Andrew, Hector's son, was a rarity of sorts, a young person still interested in working the land, fixing the ditches, and cajoling the waters. Juan Encinitas, from the Mora Valley, explained these challenges back in 2009: "None of the young people on our ditch want to work hard anymore, or they're just looking to live in the cities at this point. They're here for a couple of hours, checking their phones, then leave hours later. It's hard work, difficult to farm these days, and it's not like they're going to come back and farm full time and be able to afford a brand-new pickup and all the things that people want these days."[33] One water banker and water speculator offered a blunt prophecy a few years ago: "Acequias will be gone in 20 years."[34] I hope he's wrong because of the institutional lessons they offer us, but not for any sense of nostalgia. The ditches perform biophysical work, too: acequias keep groundwater levels high locally and essentially "bank" water locally and downstream, extending irrigation by a month to two months.[35]

Acequia advocates recognize the scalar and social limits of these local institutions. "Are you kidding?" Estevan Arellano put it bluntly in the spring of 2014. "The acequia could be really screwed up, and people sometimes fought all the time." With my students from Colorado College in tow, we walked along Estevan's dry ditch as he discussed the modern lessons of these acequias (see figure 13). These ditches are not about water equality. Parciantes are accorded water rights according to the size of their landholdings. Even so, acequias developed as logical, site-specific customary systems by which to share the actual and available water,

FIGURE 13. Colorado College students and Juan Estevan Arellano walk along the dry Acequia de la Junta y Ciénega ditch, on the lower Embudo River, in February 2014. Photo by the author.

whether in times of plenty or in times of need. Mayordomos understand annual, monthly, and weekly streamflow changes.

In hundreds of discussions over the last decade, up and down streams and ditches in New Mexico, people made it clear to me that these institutions remain important and can inform current water governance and theories about property rights.[36] The state might do well to explicitly enact proportional shortage sharing instead of the "nuclear option" of strict prior appropriation.[37] The strict ordering of seniority with prior appropriation dates, officially recognized in the New Mexico water code, is an unattractive option for both the state engineer and water users.[38] Proportional allocation, planning, and use will be vital to all states trying to coordinate interstate water sharing. The lessons from New Mexico's experience with adjudication and water settlement can advance critical legal studies and produce new outcomes that are more inclusive for those affected by the side effects of adjudication.[39]

I wrote *Unsettled Waters* because I think water rights adjudication is important and holds implications for water users, water agencies, and those of us who study water governance. So much of this legal process remains, by design, top-down. Listening to New Mexicans who value water differently produces an alternative perspective on water and water allocation. There is more to valuing water than the monetary values held in water rights. It allows experts and water users to hopefully see the value of the water concepts and governance principles practiced in this region. Sharing, equity, and participatory democratic principles abound in the institutions found in New Mexico. Planning for zero water in dry years and adjusting the shared proportion upward as water becomes available in better times seems realistic and anticipatory. As Estevan Arellano told my students back on the ditch in 2014, "We have to follow the water in our work, as it moves across the field. You coax it, but the water does not follow us." He was right. We can map water. We can award water rights. But we cannot allocate water that does not exist. In the end, we all follow the water.

INTRODUCTION

1. See Stanley Crawford's (1988) seminal description of the duties of a mayordomo (water boss) on an acequia ditch. For historical and cultural dimensions of acequias, see José Rivera (1998). Sylvia Rodriguez (2006) details the importance of these institutions for ritual tradition in New Mexico. Juan Estevan Arellano (2014) provides a powerful description of the local knowledge involved in acequia maintenance. For a good, historical overview of the significance of acequias, see Simmons (1972).

2. I am not conflating sovereignty with security (see the concerns of Gupta et al. 2016). Instead, I mobilize the concept of water sovereignty as a concept expressing the degree of water governance and control in specific territorial contexts. Local water sovereigns like acequias only control their own ditch water and have limited power beyond the ditch's reach. Pueblo reservations have greater water sovereignty that extends over the entire boundaries of the reservation, well beyond the confines or reach of the ditches they manage. The state of New Mexico claims overall governance authority over all water in the state. Finally, the federal government has nation-state claims, or "federal reserve waters," but these federal claims nestle in scale within the state boundaries since they typically apply to endangered species or represent Native/Indian water rights claims. These overlapping jurisdictional (sovereign) and spatial (territorial) claims to water sovereignty, then, explain much about why adjudication runs into problems in the state's attempt to simplify water. See Murphy (2013) for a review of why territoriality as a political process remains relevant.

3. In this book, I use Indian water rights interchangeably with Native or indigenous water rights, depending on the scale or specificity of the context.

4. *Journal of the Southwest* 32, no. 3 (1990) special volume on the adjudications of water rights in New Mexico for an early assessment of the impacts on indigenous and Hispano

New Mexicans. An unpublished dissertation by Saurí (1990) was the first to argue that local water governance was rescaled to the state engineer's office during the twentieth century. A retired state court judge also provided an interesting account of his years on the bench serving the Lower Rio Grande adjudication (Valentine 2012).

5. Levine (2008, 259). For wide-ranging critical discussions on contested water rights in law from other cultural contexts, see, for example, Roth et al. (2005) and Boelens et al. (2010).

6. See Benda-Beckmann et al. (2009), for example, on legal pluralism literature in anthropology. Aspects of scale, space, and place from geography are vital here. See the early work by Matthews (1984) and later Blomley (2003, 2008), Blomley et al. (2001), and Delaney (2003, 2010); see Bartel (2016) for a recent review of the field and Mattei and Nader (2008) for a critique of imposed law as a form for resource extraction. Roth et al. (2005) address cultural and political-legal aspects of water rights in their edited volume.

7. For representative examples, see Feller (2007), MacDonnell (2015), Thorson et al. (2005, 2006), and Tarlock (1989, 2006). Feller and Thorson address key aspects of western adjudications. The works by Tarlock and MacDonnell listed here offer powerful critiques of prior appropriation law and its potential role in adjudication outcomes.

8. For water governance, see Conca (2006). For aspects of privatization, Bakker's (2004, 2007, 2010) work remains the most nuanced analysis. For the urbanization and transformation of water into a commodity, see Swyngedouw's works (2004, 2005, 2009).

9. To cite just a few of the exemplary works, see, for example, Marc Reisner's (1986) Mormon irrigation examples in *Cadillac Desert,* which were preceded by centuries in New Mexico by Pueblo floodwater farming and the Hispano acequias. Donald Worster's (1985) *Rivers of Empire* focused on modes of production that have long coexisted side by side, not in a kind of sequential stages-of-development model. Patricia Limerick and Jason Hanson's (2012) recent and remarkably full-throated defense of Denver Water's maneuvers to move water from the mountains to the city has its own partiality toward the engineers of the past. McCool's (2002) book on Native water rights settlements is closest to what this book addresses and stands as the authoritative text on Indian water settlements. John Fleck's (2016) fine and recent volume *Water Is for Fighting Over and Other Myths about Water in the West* dedicates only two paragraphs to a minor adjudication of groundwater in Southern California.

10. Worster (1985) provides a structural critique of how water became a resource under the capitalist assumptions of western states. One recent and wide-ranging exception to this is Schmidt (2017), which focused on the intellectual and philosophical history of water in western society.

11. Following Scott's (1998) work. Adjudication is a hybrid "high-modern" state practice in that individual western states execute the process, whether administrative or judicial, based on the ideas of a federal employee, Morris Bien (in 1903). New Mexico adopted Bien's model code for water adjudications.

12. Space limitations preclude a fuller treatment here, but see Clark's (1987) exhaustive volume on differences between these kinds of water organizations in New Mexico.

13. Rodriguez (1990) was the first author I know of that leveled this critique against what was already a mature adjudication-industrial expert industry. See chapter six for more on this set of public and private political-economic concerns around water rights proceedings.

14. I do follow the predilection of historical political ecologists (see Offen 2004; Davis 2009) in sorting the experiences of the living (first) to make sense of past archival records (second). This cross-checking method between interviewee experiences of adjudication, versus the official transcripts, was especially useful for the Aamodt basin case discussed in chapter 2.

15. See Freyfogle (2003).

16. The modest hope is that this book will do for water adjudications in New Mexico what other books have done for land grants (Correia 2013; Dunbar-Ortiz 2007; Ebright 1994; Montoya 2002), forest resources (Kosek 2006), climate change (deBuys 2011), and the nuclear industry (Masco 2006). Although grounded in New Mexico, the book addresses the same dilemmas that other western states and dryland locations around the world face.

17. See one fascinating example of a public political ecology clearinghouse run by Tracey Osborne (geography, University of Arizona) at Public Political Ecology Lab, http://ppel .arizona.edu/.

18. Colorado is considered "adjudicated," although ongoing water court hearings continue. Idaho recently completed the gigantic Snake River adjudication. Montana has made remarkable progress in its own adjudication but remains incomplete. See Bryan (2015) for more on the status and nature of the various state adjudications in the American West.

19. I am not arguing that all state visions projects produce negative consequences. See Agrawal (2005) for aspects of citizen self-monitoring that produce new forms of environmental citizenship. Similar by-products have occurred during and after adjudication, as I demonstrate in chapter 7. See also Carter (2008, 2012) regarding Argentina's progressive-era efforts to control malaria.

CHAPTER 1. HOW LOCAL WATERS BECOME STATE WATER

1. M. Otero, personal interview, August 3, 2011, Rio Lucio, near Picuris Pueblo, NM.

2. See Nostrand (1992) for more on the shifting labels of Hispano, Spanish American, etc.

3. See Rivera (1998) for more on the historic roots of acequias and Wescoat (1995) for more on the Moorish law of thirst. In the state's water code, there is even a provision reflective of this influence for protecting travelers and their access to water in times of need. See also Clark (1987, 9).

4. See Saranillio (2015) for an excellent summary on *settler colonialism*. Seminal works by Wolfe (2006) and Veracini (2010) further explain this concept. The consequences of two separate waves of settler colonialism in New Mexico are summarized in Wilmsen (2007).

5. See Correia (2013) for more on the dispossession of territory and land grants.

6. See, for example, Roybal's (2017) work on gendered property dispossession during the territorial period of New Mexico.

7. Benton, delivered on the floor of the US Senate, January 15, 1849. AYER 160.5 .C15 B47 1849 – Special Collections, Newberry Library, Chicago, IL.

8. An alternate, less generous reading of Benton's objections might also include his reluctance to include New Mexicans into the Unites States.

9. See Conca (2006) for more on water governance. Sovereignty as a concept is powerful because it also includes the notions of territorial control, or territoriality, as a concept.

Water sovereigns are simultaneously worried about decision-making (governance) and the loss of land and water integrity and continuity (territoriality).

10. Scott (1998).

11. This is adapted from Maria Montoya's (2002) excellent work on the Maxwell land grant and how notions of law between US and Spanish forms of landed property were poorly translated.

12. Loss of Spanish Mexican-era land grants and forest access is detailed by Montoya (2002), Kosek (2006), and Correia (2013).

13. The water right must be registered already with OSE or be registered at the point of selling the water right, via documentation in an administrative hearing. Individual water use rights do not depend on a full adjudication in order to be sold.

14. For more on neoliberal natures (nature under late capitalism), see Wilder and Lankao (2006), Heynen et al. (2007), Sangameswaran (2009), Lave (2012), and Budds (2013). Most of these, along with the work by Whitehead et al. (2007), focus on the nation-state (federal) aspects of neoliberalism. Adjudication is a state-led process that parallels the concerns of Scott (1998) and Ferguson and Gupta (2002). Capitalism served through law as a frame combines the insights of Mattei and Nader (2008) with Harvey (2003), in that "accumulation by dispossession" of water can be done through law or by ignoring it (see chapter 8).

15. See Meinzen-Dick and Mwangi (2008) for a recent example of some of the pitfalls in formalizing property rights and how using individual (and usually male) bodies as recipients of these new property rights can exclude access to women, children, and the elderly.

16. See Thorson's (2001) assessment on western adjudications, especially appendix B, a table that lays out the various western state efforts and distinctive water codes with numbers of parties involved (as of 2001). See also the work of Alatout (2007, 2008) for comparative work on identity and the making of state water.

17. D.L. Sanders, "Adjudication: Curse or Salvation." Comments delivered to the 10th Annual Assembly of the New Mexico Water Dialogue, June 10, 2006, Dane Smith Hall, University of New Mexico, p. 5.

18. Bryan (2015, 487) estimates that less than half of all water rights are decreed. What this means is that individual subfiles (water rights) are better understood, even if the total surface area of basins lags behind, because in New Mexico full basins are only decreed when all stages of the process are finished.

19. This is a key distinction to understand: adjudications stay in the courts, through formal litigation. Water agreements, or "settlements," are taken out of court by the interested parties in a water rights lawsuit but often duplicate many of the same procedures.

20. To paraphrase Scott (2009), they pursue through settlement what one might call one of the many arts of not being governed (by the state).

21. See Thorson et al. (2005, 2006) for more on western water adjudications. See also Tarlock (1989, 2006) on water adjudications in the western United States. Feller (2007) discussed how adjudication in Arizona severely affected that state's response to legal water challenges. Clark's (1987) volume does discuss adjudications but privileges archival and expert views of water. Pease (2010) and Shively (2001) examined how nonadjudicated waters complicate water transfers in the Middle Rio Grande. Earlier works by the author (Perramond 2012, 2013) and with Maria Lane (Perramond and Lane 2014) discuss some of

adjudication's *effects* in New Mexico. The Snake River adjudications in Idaho offer insights on how to make adjudication in New Mexico more efficient. See Jones (2016).

22. E. Cornud, personal communication, July 18, 2011, Embudo, NM.

23. See McGranahan (2016) for more on the "refusal of the state."

24. For more on federal irrigation policy and its history, see Pisani (2002). This is not the first software analogy for water; see Sedlak (2014) for more.

25. Note that "native" water is a designation for water from its basin of origin. I do not mean "Indian water" in this case. Water from the Chama, for example, contains San Juan (Colorado River basin) water before it enters the Rio Grande "native" flows.

26. See Meehan (2014) for more on the tool power of water infrastructure and how it not only shapes behaviors but has larger repercussions for state power.

27. L. Ranger, personal interview, September 30, 2009, Santa Fe, NM.

28. T. Orviedo, personal interview, November 8, 2009, Taos, NM.

29. T. Robinson, personal interview, February 2, 2009, Santa Fe, NM.

30. Prior appropriation often disincentives any water conservation measures, like rain barrel collection. See Meehan and Moore (2014) for one recent example of the consequences of this.

31. Despite the seeming attractiveness of watershed logic and management (e.g., Worster 2003), there are real challenges to watershed management and whether it actually simplifies the complex cultures and politics of water use. See Molle (2009) for more on watershed management and Blomquist and Schlager (2005) and Cohen and Davidson (2010) for the complications with a watershed approach.

32. Following Barnes's (2014) treatment of water in Egypt, cultivators and their practices are vital to understand when it comes to water use and water resources development. See also Blomley (2013) regarding the notion of performing property and Correia (2013) for New Mexican instances of property performance through land tenure claims. An acequia holding their annual meeting as an institution, to agree on watering schedules and timing, is an instance of "performing governance" (Robertson 2010).

33. See Correia (2013) for the Spanish Mexican-era landed property possession acts.

34. L. Martinez, personal interview, November 9, 2009, Taos, NM.

35. B. Holman, personal communication, August 12, 2012, near Mimbres, NM.

36. This binary approach to indigenous identity is complex. For more on Indian "blood quantum" rules, see Hill and Ratteree's (2017) excellent edited volume. See also TallBear's (2013) work on the dangers of how DNA can put sovereign claims at risk. Sovereign nations, then, should be the entities that "claim" who should belong and be enrolled in tribal registers and rolls. The Indian binary remains problematic for those who claim joint heritage identity as Indo Hispano or those who stake claims as members of Genízaro (Christianized Indian) villages in New Mexico.

37. See Barrett (2002) in general for Spanish settlement and Mathews-Lamb (2001) for a more detailed, historiographic account of identity, history, and water claims in this area.

38. See Brooks (2002) for more on these complicated kinship networks in colonial New Mexico.

39. See the work by Kauanui (2008) on blood drop rules in Hawai`i and how these federal recognitions cleave identity without necessarily affirming sovereignty for indigenous

peoples. Barker (2011) argued that US federal inscriptions of authenticity need to be questioned by sovereign indigenous nations.

40. See Coulthard (2014) for an incisive critique about the contentious work of nation-state "recognition" and how these often preserve colonial mind-sets and politics.

41. Gupta et al. (2016) are correct in cautioning scholars who might conflate "water sovereignty" with "water security." Here, I use water sovereignty as a nested concept with multiple levels of power: with acequias, indigenous reservations, the state of New Mexico, and the US federal government holding increasing levels of water sovereignty.

42. There are technically nineteen Indian pueblos and three Indian reservations in New Mexico. The three Indian reservations are the Navajo (by far the largest), the Jicarilla Apache in northern New Mexico, and the Mescalero Apache in southern New Mexico. The Pueblos, under the Spanish, Mexican, and treaty rights of Guadalupe-Hidalgo (1848), were awarded "Pueblo league" grants under the law. US assignments of later Indian reservations were conducted separately.

43. Boelens (2009) has written one of the few critical pieces regarding how "water rights" as a framework individually disciplines our conceptions of water as a property severable to individual bodies. See also Boelens et al. (2010) for the best single edited volume addressing individualized water rights.

44. See Bakker (2007) for these important distinctions. Privatizing water is not the same as making water a commodity. The private use right in the western United States is a limited right constricted by the state's claims that all waters belong to the public.

45. See Blomley (2008) on how the simplification of water and land issues as property is never simple. Bureaucracies and expert agencies, as they try to simplify water, also become dependent on the users they try to govern (see Banister 2014 for a parallel in Mexico).

46. See Arellano (2014, 199).

47. Jerald Valentine (2012, 50) included this phrase of "litigation illuminates." He noted that the "goal of a just, speedy, and inexpensive determination of water rights in adjudications remains elusive" (51). This essay is in Ortega Klett's (2012) edited volume on water in New Mexico.

CHAPTER 2. AAMODT, DAMMIT! BIG TROUBLE
IN A SMALL BASIN

1. A. Borique, personal interview, July 28, 2015, Pojoaque, NM.

2. I chose the cases described in this and the following chapter because they exemplify aspects that make adjudication difficult. These cases influenced others, as did their eventual settlements.

3. Colonial-era water disputes in the Pojoaque are well documented in Baxter (1997).

4. T. Adel, personal interview, October 29, 2009, Taos, NM.

5. Two different judges ruled on motions of the case and thus legal scholars often refer to Aamodt 1 and Aamodt 2 based on the different judges and rulings.

6. P. Robinson, personal interview, August 3, 2011, Santa Fe, NM.

7. See Doolittle (2000) for a summary of Native agricultural practices in the region.

8. T. Duran, personal interview, July 28, 2011, Santa Fe, NM.

9. See Blair (2017) for more on this concept of settler indigeneity and its problematic implications.

10. For this summary of the Aamodt 1 and Aamodt 2 distinctions in the case, I draw heavily on Peter Chestnut's document written for *WaterMatters!* 2013, part of the University of New Mexico Law School's Schell Ombudsman Program, http://uttoncenter.unm.edu /pdfs/Water-Matters-2013/Aamodt%20Adjudication%20.pdf.

11. See Vlasich (2005, 164) for more on the poor market value awarded to the Pueblos for lost water rights in 1924. See also Vlasich (2005, 282) for more on the Mechem decision that awarded only historical irrigated acreage water rights to the four pueblos; one Nambé Pueblo parcel did get "reserved" waters instead of historical rights due to a later award of those lands.

12. S. Schillings, personal email communication, November 30, 2011.

13. P. Lessing, personal interview, November 9, 2009, Santa Fe, NM.

14. C. Ortiz, personal interview, July 28, 2012, Santa Fe, NM.

15. W. Anderson, personal interview, August 15, 2011, Santa Fe, NM. This was a common refrain and fear: that well owners would be seriously curtailed. Since the groundwater and surface water are so closely related, one can also understand the Pueblos' concerns about these domestic wells.

16. For a sample of this, see the *Santa Fe Reporter* story dated August 24, 1983, titled "The Water Lawsuit's Bitter Taste" by Stephen W. Terrell, Fray Angélico Chávez Historical Archives, Collection AC482, Aamodt Water Case, box 1, Santa Fe, NM.

17. D. Martin, personal interview, November 14, 2009, Tesuque, NM.

18. These are the real names of the attorneys, as part of the official proceedings cited from the Aamodt case collections in the Fray Angélico Chávez Historical Archives; these lawyers then represented most of the non-Indian defendants in the Pojoaque Valley. The remaining names from more recent interviews in this chapter are, as usual, pseudonyms to protect identity.

19. Fray Angélico Chávez Historical Archives, Collection AC482, Aamodt Water Case, box 1, fol. 1, "Pojoaque Valley Water Users Association, " Santa Fe, NM—letter (confidential communication) dated August 5, 1985, from Peter B. Shoenfeld.

20. On the "legal slush fund"—see Fray Angélico Chávez Historical Archives, Collection AC482, Aamodt Water Case, box 1, fol. 1, "Pojoaque Valley Water Users Association," Santa Fe, NM—memo to members, undated document, 1985, written by D. Ortiz.

21. W. Anderson, personal interview, August 15, 2011, Santa Fe, NM.

22. See Baxter (1997), which describes the long-held customary sharing and allocation of water in northern New Mexico and explores many of the more infamous Indo-Hispano conflicts and legal cases over water rights.

23. These accounts are from Terrell, "Water Lawsuit's Bitter Taste."

24. M. Inerque, personal interview, November 29, 2010, Santa Fe, NM.

25. O. Gomez, personal interview, August 1, 2011, Pojoaque, NM.

26. E. Romero, personal communication, November 6, 2009, Albuquerque, NM.

27. J. Freschett, personal communication, November 19, 2016, Taos, NM.

28. As State Engineer Steve Reynolds said in 1983, "Certainly this has continued longer than I had anticipated. But I was aware that would involve adjudication of pueblo Indian water rights and that this matter might have the effect of protracting the litigation. I would have been a fool if I had not, wouldn't I?" In the same story, one can see the firmness of opinion that Reynolds had and that his coworkers reported of him, "Once the matter is

settled in an adjudication . . . that's the end of the road." *Santa Fe New Mexican,* B3, September 18, 1983, special issue, Fray Angélico Chávez Historical Archives, Collection AC482 Aamodt Water Case, box 1, Santa Fe, NM.

29. A. Roybal, personal interview, January 16, 2010, Santa Fe, NM.

30. See, for example, a recent story on how the new regional water system will roll out and valley resident concerns about it at "New Water System in Works as Part of Aamodt Settlement," *Santa Fe New Mexican,*" February 11, 2017, www.santafenewmexican.com /news/local_news/new-water-system-in-works-as-part-of-aamodt-settlement/article _6a876c71-56cc-5054-b61e-7b2586ceb121.html.

31. In chapter 4, I address some of the biophysical and legal consequences of the Aamodt settlement. This quote is hinting at the 1982 ruling that forces any post-1982 domestic well use into a low withdrawal rate assignment to reduced use (between 0.3 AFY to 0.7 AFY).

32. T. Robinson, personal interview, July 28, 2015, Taos, NM.

33. S. Baca, personal interview, July 28, 2015, Taos, NM.

34. See Baxter (1997) for a variety of cases in which the Pueblos filed encroachment complaints against their lay Spanish neighbors during the colonial period.

CHAPTER 3. ABEYTA: TAOS STRUGGLES, THEN NEGOTIATES

1. E. Mondragón, personal interview, November 15, 2009, Taos, NM.

2. R. Dempsey, statement in public meetings on July 27, 2015, Taos, NM.

3. Taos Regional Water Plan, 2016. See page ES-1 in their executive summary for the statistic on surface water, as well as the plan for any drought-induced shortages.

4. See Cox (2014) regarding the internal struggles of the Taos Valley acequias.

5. See also Espeland's (1998) excellent treatment on how the Arizona Pima stopped the development of a dam in Central Arizona. More on the Indian Camp Dam episode is found in John Nichols's direct and witty assessment in Ortega Klett (2012, 166–179).

6. B. Thompson, personal interview, August 20, 2011, Taos, NM.

7. Martin Reuss (2008) has documented how much of actual "engineering" has incorporated this growing element of public negotiation when it comes to civil engineering, especially, and how "seeing like an engineer" has morphed into a more transparent process, especially in planning processes for major works.

8. M. Asturias, personal interview, November 11, 2009, Ranchitos de Taos, NM.

9. N. Tafoya, personal communication, February 1, 2010, Talpa, NM.

10. T. Domingo, personal interview, February 1, 2010, Talpa, NM.

11. J. Robinson, personal interview, February 3, 2010, Santa Fe, NM.

12. A. Gurulé, personal interview, January 21, 2010, Taos, NM.

13. P. Martinez was a key figure in Rodriguez's (2006) work, and this is not a pseudonym. See Chief et al. (2016) for more on these key practices to get engagement between cultures of water in the greater Southwest.

14. Rodriguez (2006, 126).

15. O. Ramirez, personal interview, January 23, 2010, Taos, NM.

16. A partial final decree for the Rio Truchas drainage was entered in 1974.

17. W. Merland, personal communication via email, November 3, 2010. This conclusion might be questionable, since new settlements and interstate haggling do often result in par-

ties and states being able to further develop water resources. The 2004 Arizona settlement act allows New Mexico the use of the Gila River, some 14,000 AFY, as part of that settlement.

18. Nichols, in Ortega Klett (2012, p. 173).

19. The first part translates to "water is life."

20. Rodriguez (2006) makes this point clearly throughout her book.

21. Alfonso Martinez, interview with the author, February 1, 2010, Taos, NM.

22. V. Gallegos, personal interview, November 11, 2009, Taos, NM.

23. G. Suazo, statement during New Mexico legislature Water and Natural Resources Committee meeting, July 27, 2015, Taos, NM.

24. When asked about this quote after the meeting, the state senator in question asked for anonymity.

25. It passed Congress as H.R. 5122 during the 102nd session of Congress in 1992.

26. Taken from H.R. 5122 language, specified to the Jicarilla Apache tribe. This tailored approach was necessary given the multiple groups that share the same streams—in this case the nearby Navajo claims that were only settled in 2005 between the Navajo Nation and the state of New Mexico.

27. D. Ortiz, personal interview with the author, February 1, 2010, Taos, NM.

28. T. Rodriguez, personal communication, February 1, 2010, Taos, NM.

29. McCool (1987).

30. McCool (2002, 46).

31. G. Valdez, personal interview, February 2, 2010, Taos, NM.

32. To reiterate: there is no physical hydrologic connection between the Chama and the Taos area, thus Taos Pueblo has been awarded "water" that flows elsewhere (from the project) yet will be physically pulled from the local valley. It is a legal and political trade-off and not a biophysical connection to real water from the Chama.

33. G. Valdez, personal interview, February 2, 2010, Taos, NM.

CHAPTER 4. LOCAL SETTLEMENTS CONNECT WHAT STATE ADJUDICATION SEVERED

1. Geraldine Moya-Gurulé, personal communication, July 26, 2015, Taos, NM.

2. This is especially true for the unresolved Indian water claims in New Mexico. See Hughes's (2017) recent summary of the implications of unquantified Pueblo Indian claims for the Rio Grande.

3. Taken from McCool (2002, 58).

4. See Titus (2005) for more on the challenge of domestic wells in New Mexico.

5. As a reminder, these are the Nambé, Pojoaque, Tesuque, and San Ildefonso Pueblos.

6. The proposed location, as of 2018, for the new Pojoaque (Aamodt) water regional system is set north of the gauge, avoiding this potential river compact issue. Readers should note that the United States Geological Survey formally uses the spelling "gage" over "gauge." See chapter 5 for more on this.

7. See, for example, the story by Kay Mathews from 2016 on how a water transfer from the Questa area north of Taos was built into the terms of the Abeyta settlement. "Taos Acequia Still Objects to Terms of Abeyta Settlement," *La Jicarita*, July 24, 2016, https://laji-carita.wordpress.com/2016/07/14/taos-acequia-still-objects-to-terms-of-abeyta-settlement/.

8. See "Top of the World Water Rights Transfer Spurs Worries, Plans," *Santa Fe New Mexican,* July 4, 2015, www.santafenewmexican.com/news/local_news/top-of-the-world-water-rights-transfer-spurs-worries-plans/article_d457d2dc-4de7-5f8″3-bc2f-539aa8f3a831.html for a news story that illustrates the rhetorical presumption that the Top of the World Farms transfer would be approved by the OSE because of the political pressure on settling both basins.

9. Kay Mathews has been particularly attentive to this lingering, controversial water transfer. "Taos County Waits to Hear Whether Top of the World Water Will Be on Its Way South," *La Jicarita,* November 14, 2016, https://lajicarita.wordpress.com/2016/11/14/taos-county-waits-to-hear-whether-top-of-the-world-water-will-be-on-its-way-south/.

10. D. White, personal interview, February 5, 2010, Santa Fe, NM.

11. E. Robinson, personal interview, August 2011, Santa Fe, NM.

12. See "Federal Judge in Aamodt Case Recuses Herself," *Santa Fe New Mexican,* July 15, 2014, http://www.santafenewmexican.com/federal-judge-in-aamodt-recuses-self/article_21a2f52e-5412-5d6e-89ef-5ad612510f64.html for the recusal of Judge Martha Vázquez on July 15, 2014, from the Aamodt settlement.

13. S. Enders-Gomez, personal communication, July 26, 2015, Taos, NM.

14. The September 2017 deadline was met, meeting the terms embedded in the Aamodt and Abeyta settlements.

15. Decades ago, Bowden (1977) made this case. Glennon (2002) emphasized Bowden's message with a wider survey of the effects of groundwater pumping.

16. See the "Taos Water 101: The Abeyta Settlement Explained," *Taos News,* July 26, 2012, www.taosnews.com/stories/taos-water-101-the-abeyta-settlement-explained,6563 for one example of local Taos reporting that summarizes the complex ground and surface water mix involved in the settlement.

17. T. Evans, consulting hydrologist, personal communication, August 15, 2014, Santa Fe, NM.

18. T. Moran-Durazno, personal communication, July 28, 2015, Taos, NM.

19. T. Herrera, personal communication, August 2, 2014, Santa Fe, NM.

20. These are called post-1982 wells, even if the signed moratorium on new domestic wells was ordered by the court on January 13, 1983.

21. B. Riley and J. Saunders, personal communication, August 7, 2011, Santa Fe, NM.

22. J. A. Blair (2017) used this concept of *settler indigeneity* in the Falklands (Malvinas) to describe how long-term (Anglo) residents claimed Native identity by performing environmental management acts on the islands. In New Mexico, Hispanos on acequias make similar arguments for settler indigeneity claims.

23. B. Barrett, personal interview, September 28, 2009, Albuquerque, NM.

24. J. Heard, personal communication, February 1, 2010, Taos, New Mexico.

25. See Price (2011, 77).

26. These are the latest available estimates for the costs of the two (Aamodt, Abeyta) settlements. Sourced from New Mexico OSE documents.

27. Engineering is critical to the settlement terms throughout the entire Pojoaque Valley system, with some of that water putatively being transferred from the Abeyta case (Taos Valley water). This highlights new roles for engineers dragged into the negotiation process (Reuss 2008).

28. See McCool (1987) for more on the original configuration of iron triangles in the western states.

29. As settlements create new infrastructure, these will also alter river channels in New Mexico. More attention to these legally modified physical geographies would be wise (Lave 2015).

30. For historical context, see Meyer (1984) and Baxter (1997). Adjudications simply resurfaced these water struggles in a new way (see Levine 1990; Quintana 1990; Rodriguez 1990, 2006).

31. McCool (2002, 23). See also Brienza (1992) on settlements and DeJong (2016) on the tragic Pima example in Arizona.

CHAPTER 5. CHANGING MEASURES: HOW EXPERT METRICS CHANGE WATER

1. Juan Estevan Arellano, interview with the author, October 27, 2009, Embudo, NM.

2. See Carroll (2006, 2012) and Akhter (2016) for discussions of the "data state" and technocratic water expertise. Paradoxically, the more experts and agencies try to depoliticize water through "better data" and public participation, the more the opposite happens: water becomes increasingly political. Such is the case with adjudication in New Mexico.

3. A note for readers on spelling: the USGS recognizes *gage* instead of the more accepted *gauge* because of Newell's preference for the spelling of the word (favoring *gage*). Newell, the founder of the Embudo Station "camp of instruction" for hydrographers, oddly thought the *gauge* spelling was tainted by Saxon influence and systematically used *gage* thereafter. Here I use *gauge* when referring to the devices abstractly and use *gage* when referring to the USGS devices specifically. For more on Newell's strange convention and the USGS spelling, see the technical report by Follansbee (1919), https://pubs.er.usgs.gov/publication/7000087.

4. For more on Powell, Dutton, and Newell, see Worster's (2002) magisterial work on Powell.

5. Linton (2008, 2010) compellingly describes how water as substance became a scientific compound and object for abstraction. The links between expertise and statecraft and water's link to colonialism are older; see Mitchell (2002). An excellent analysis of courts, expertise, and the rise of the New Mexico state engineer's role is provided by Lane (2011). Livingstone (2003) provided the clearest argument for understanding how science and expertise were dependent on particular places.

6. Taken from a local press story about the commemorative event, see the *Grant County Beat*, April 21, 2014, http://www.grantcountybeat.com/news/non-local-news-releases/15380-125th-anniversary-celebration-of-first-usgs-streamgage-located-in-embudo-nm.

7. See Aguilar-Robledo (2009) for more on the general decline in surveying throughout New Spain. Not only was there great variability in the units and measures practiced, there was also an enduring tradition of customary measures that made any standardization difficult.

8. See Baxter's (2000) article on how slippery the unit of surco was and remains.

9. For more on acequia terminology and landscape taxonomy, see Arellano (2014).

10. See, for example, Skaggs et al. (2011) and Samani et al. (2013) for a better estimate of the actual crop consumptive use of water as determined with the help of remote sensing. Samani et al. (2013) also warn about the full water duty allocations being awarded in the

Lower Rio Grande and how these might endanger Rio Grande Compact obligations and groundwater resources.

11. M. Magañez-Smith, personal interview, July 23, 2013, Pojoaque Valley, NM.

12. See Bryan (2015) for an overview of state adjudication practices and how New Mexico is pioneering in its inclusion of consumptive use during adjudication.

13. For a clear explanation about consumptive use, versus total water allocations, see Jones and Cech (2009, pp. 106–116). Ditch associations, like acequias, have more ability to contest water sales and transfers because they are more likely to be affected by adverse changes to ditch flow should water leave their ditch system.

14. A. Vorán, personal interview, August 15, 2012, Las Vegas, NM.

15. As recounted by A. Vorán, Las Vegas, NM.

16. A. Chávez, personal interview, August 2, 2011, Mimbres, NM.

17. The links between institutional practices in agencies and their use of drought moments to change water governance has been noted by many (see Hess et al. 2015 for a recent example). On a pragmatic level, changes to water management under climate change will be needed to lessen water conflict (McKinney and Thorson 2015).

18. Crawford (2003, 48).

19. B. Reedle, personal communication, August 3, 2011, Mimbres, NM.

20. See Tarlock's (2010) article on whether western water law (especially prior appropriation) can adjust accordingly with reduced water flows under global climate change.

21. L. Hernández, personal interview, August 3, 2011, Mimbres, NM.

22. A. Gabaldón, personal interview, August 2, 2011, Mimbres, NM.

23. T. Ruiz, personal communication, August 11, 2011, San Lorenzo, NM. These contemporary ethnographic allegations were not mere hearsay. As careful reading in the UNM Center for the Southwest special collections makes apparent, these accusations were pretty common in the late 1980s. See, for example, Tonantzin Land Institute Records, 1911–2000, boxes 1–2, MSS 666 BC, box 1, fol. 5, "Acequias del Valle Rio Mimbres," for an exchange alleging rude, racist treatment between one local mayordomo and OSE personnel in the Deming, New Mexico, office. Three other informants in 2009–2012 supported this allegation. Relations are said to have improved since then.

24. Interestingly, a nonmarket view of acequias yields a different appreciation for the value of these ditches (see Archambault and Ulibarri 2007).

25. D. Freed, personal phone conversation, August 2013.

26. On distinctions between privatizing and commoditizing water, see Bakker (2004, 2010). Swyngedouw (2005) described why water privatization is culturally contentious. Public municipal water systems have their challenges and are no panacea (see Zetland 2011).

27. From Reno-Trujillo et al., 2013.

28. As Meinzen-Dick (2007, 15202–15203) wrote, "Attempts to introduce formalized water markets are often met with objections to the privatization and commercialization of water because of norms that water is a free good or a gift from god . . . water markets cannot be transplanted without due consideration for their fit within the physical, institutional, and cultural environment."

29. Ibid, p. 5.

30. A. Velarde, personal interview, August 15, 2011, near San Lorenzo (Mimbres), NM.

31. See Conca (2006) for more, and see also Sedlak (2014).

32. C. Valenzuela, personal interview, August 9, 2011, near Mimbres, NM.

33. Mitchell (2002) and Birkenholtz (2008) have highlighted these problems for Egypt and India, respectively.

34. Zetland 2011, 217. See also Colby et al. (1989) and Matthews (2003) for the legal and institutional challenges of moving water through markets. Bauer (2004) has discussed how Chile reformed its water code to encourage a private water market.

35. See Brookshire et al. (2013) for a useful compendium on water policies in New Mexico that offers traction for these problems.

36. See Shiveley (2001) and Pease (2010) for more on the growing trend but difficult nature of tracking water rights sales in New Mexico. Like many aspects that New Mexicans complain about regarding water, the availability and transparency of these data are questionable.

37 . This is a chronic problem in New Mexico. See Brown and Ingram (1987), Ingram and Brown (1987), Jepson (2012) for a Texan context and example, and Wescoat et al. (2007) for a broader southwestern perspective. A recent edited volume by Whiteley et al. (2008) discusses water equity in several cultural contexts.

38. See Zetland (2011, 2014) for some creative solutions using market principles. See also Brewer et al. (2008) for more on the challenges and promises of water markets and trading in the western states.

39. A. Gabaldón, personal interview, August 2, 2011, Mimbres, NM.

40 . This was a repeated sore point in at least thirty-nine discussions with local water users; namely, that the state presumed to own all the water and then to "give it back" to the people who preexisted the actual state (following the thoughts of Boelens 2009; Parenti 2015).

41. Bauer (2010) provides comparative perspective for western water market proponents based on his work in Chile.

CHAPTER 6. WORKING FOR THE ADJUDICATION-INDUSTRIAL
COMPLEX

1. Taken from Mitchell (2002, 15).

2. There is no shortage of comparisons available here. In Chile, for example, Budds (2009) documented how often the national water agency uses a whole variety of consultants to create the appearance of objectivity even when most consultancy reports support the premise and desires of the water agency. "As the neoliberal system privileges technical expertise, it offers no scope for non-specialist contributions in decision-making processes, which could largely explain why the assessment was entirely top-down with no inclusion of public participation or local knowledge" (427). Like the OSE in New Mexico, the water department, DGA (Dirección General de Aguas), in Chile has to walk a fine line between respecting the democratic liberal rights that flow from the Water Code and trying to retain the "decision-making power denied to it under the Water Code and the wider political economic regime" (427).

3. Dan Tarlock (2006) has likewise questioned the overall "worth" of general stream adjudications and whether they have produced anything positive. He notes that Montana had spent nearly forty million dollars on a single-stream adjudication (as of 2005) but also warned that actually getting figures on the cost-benefit of adjudication was tricky.

4. Costs were calculated as follows: Total settlements from US federal commitments, state commitments for New Mexico, and costs of ongoing adjudication only for the last thirteen years from the Ridgeley (2010) public presentation based on an annual salary budget for litigation and an adjudication program budget on the order of five million dollars a year (2010 budget figures). The costs estimated here do not include the vast amount spent on or from the private sector or any of the infrastructure involved before 1980 or to be resolved in the future. Several emails in December 2016 and January 2017 to verify these costs with OSE went unanswered.

5. S. Enderton, personal interview, August 8, 2011, Santa Fe, NM.

6. Taken from the *First Biennial Report of the Territorial Engineer* (1908, 28).

7. See Tarlock (1989) for more on Bien and the development of a model water code.

8. Some would call this a neoliberalization of state practices, but this terminology is not perfectly aligned with hollowing out the state. Here, the state of New Mexico depended on federal resources to accomplish early work called for in its state water code. Certainly, the combination of public state and federal funds have benefited private industry across the board as adjudication and later water settlements "rolled out."

9. From Jones and Cech (2009, 227).

10. R. Nelson, personal interview, Taos, NM, October 11, 2009.

11. This particular deBuys quote comes from *TomDispatch.com*, July 30, 2013, http://www .tomdispatch.com/blog/175730/, but see deBuys (2011) on climate change in the Southwest. This joke has become its own bad parody, as any controversial new water measure introduced in the state legislature is also referred to as a "perpetual lawyer employment guarantee."

12. Just as new forms of technology beget new and future forms of technology, new legal statutes and modifications typically generate more employment and future cases for attorneys.

13. T. Gallegos, personal interview, July 26, 2015, Taos, NM.

14. R. McLennon, personal interview, January 15, 2010, Santa Fe, NM.

15. Readers can currently find the op-ed written by Senator Ortiz y Pino at Senate Democrats, 2015, http://www.nmsenate.com/2015/07/27/we-need-wet-water-not-paper-water/.

16. L. Ortiz, personal communication, August 8, 2011, Mimbres, NM.

17. R. Thompson, personal interview, October 28, 2009, Taos, NM.

18. This particular figure and example comes from sheet map 12 of the Taos hydrographic survey, http://www.ose.state.nm.us/HydroSurvey/RioPueblo-deTaos/rgta-12.pdf.

19. See Bryan (2015) on this point. Most states do not keep their adjudication data updated.

20. See Harley (1988) for more on power and cartography and their essential function to crafting notions of spatial power. Cartography has been the handmaiden to colonialism, imperialism, and nationalism for as long as maps have been created.

21. These hydrographic surveys calculated consumptive use, not just acre-feet of water duty per crop, which is an advantage for New Mexico compared to most other western states, as discussed in the previous chapter. What remains in error are the changes in cropping, ownership, subdivision of property, and how much crop land may have been retired from farming since the time of the survey (Bryan 2015).

22. B. Jenkins, personal interview, September 23, 2009, Santa Fe, NM.

23. T. Travis, personal interview, February 2, 2010, Santa Fe, NM.

24. B. Somme, the engineer at the beginning of the chapter, personal communication with the author, August 18, 2009, Santa Fe, NM. Quote taken on a 2009 visit to what would

become the now extant Buckman Direct Diversion plan that provides Santa Fe with additional water.

25. B. Ferguson, former Bureau of Reclamation engineer, December 11, 2010, Albuquerque, NM.

26. L. Davidson, personal interview, December 5, 2009, Santa Fe, NM.U.

27. The largely aspirational plan for Colorado can be found at Colorado's Water Plan, n.d., https://www.colorado.gov/pacific/cowaterplan/plan.

28. As noted by Conca (2015), however, the rise of integrated water resource systems (like AWRM measures in New Mexico) has simply led to a further call for infrastructure that is more about piping than about dams.

29. Hall (2002, 1) remarks on this term in his book on the Pecos River court struggle between New Mexico and Texas.

30. W. Smithson, contract historian, personal communication, November 8, 2009, Santa Fe, NM.

31. See Rodriguez (1990, 2006) for examples of anthropological and archaeological contributions to courtroom expert testimonies and the productive role they can play in shaping adjudications and court findings.

32. J. Taylor, retired historian, personal communication, November 11, 2009, Santa Fe, NM.

33. See Tyler (1990) and Meyer (1984) for guides on previous Spanish and Mexican water principles and laws.

34. For example, see Benson's (2012) analysis of three different state cases that put prior appropriation principles into question. He goes on to say that "the western states have been quietly moving away from PA for many years, abandoning it in stages" (709).

35. K. Matthews, *La Jicarita*, June 28, 2015, https://lajicarita.wordpress.com/2015/06/28/new-mexico-water-adjudications-oh-what-a-tangled-web-we-weave/.

36. C. Long, personal communication, January 21, 2010, Santa Fe, NM.

37. T. Albrecht, former OSE adjudicator, November 30, 2009, Santa Fe, NM.

38. One of the finest dark humor passages from the Gila River adjudication in Arizona can be found in Feller (2007): "[One] does not 'get out' of the Gila adjudication. It is a sort of judicial black hole into which light, sound, lawyers, water—even Judge Goodfarb—indeed, whole forests of paper, will disappear. The only way out is out the other end." Describing adjudication as a "black hole" is not isolated to this instance, either, as many New Mexicans used this exact phrasing to describe their own experience with the process.

39. This phrasing owes much to the work of Ruth Meinzen-Dick (2007). As she stated, "Program developers and implements to be able to recognize existing conditions and how they vary within a country, state, or even a single project, instead of 'seeing like a state' and expecting uniformity across all sites. A particular challenge is for state agencies to identify the existing degrees of organization among users and the institutional bases for cooperation. Too often, only registered organizations are visible to the state agencies" (15205).

CHAPTER 7. NEW WATER AGENTS AND ACTORS IN
CIVIL SOCIETY

1. See E. Wolf (1982, x).

2. N. Garcia, personal interview, August 2, 2011, Santa Fe, NM.

3. There are striking similarities here to what Agrawal (2005) documented in his book *Environmentality* regarding how particular technologies of government are adopted and then self-enforced within civil society groups interested in monitoring natural resource use. In New Mexico, Chicano (and Pueblo) notions of water push back against this grain of individual motivations for treating water-as-property (see Peña 1999 for one example).

4. See Dawson et al. (2014) for more on state territoriality across all levels of governance. The state's notions of territoriality are long-rooted in cadastral land biases (Wainwright and Robertson 2003) yet extended into water during the last century (Carroll 2006). See Hannah (2000) for discussions of nineteenth-century landed forms of territoriality in the western United States, Ingram (1990) on the recruitment of water into state-based politics, and Parenti (2015) and Murphy (2013) on how territoriality continues to linger in its logic.

5. V. Ordaz, personal interview, November 9, 2009, Santa Fe, NM.

6. On September 30, 2017, the US District Court for the District of New Mexico issued a memorandum opinion and order holding that the Pueblos of Santa Ana, Zia, and Jemez do not have Aboriginal water rights. Memorandum Opinion and Order Overruling Objections to Proposed Findings and Recommended Disposition Regarding Issues 1 and 2, Doc. 4397, United States on behalf of Pueblos of Jemez v. Abousleman. The memorandum opinion is the latest in the adjudication battleground over Pueblo water rights.

7. See Hughes (2017) on this aspect of the Jemez/Abousleman adjudication. See also Miller (2003) for more on the decline of using PIA standards for awarding Indians Winters doctrine water rights.

8. T. Simpson, personal interview, August 2, 2012, Santa Barbara, NM.

9. E. Avaro, mayordomo, personal interview, August 13, 2011, Alcalde, NM. All quotes from Ellen hereafter refer to this discussion.

10. The expression "dancing for water" during adjudication is from Crawford (1990). Similarly, see Robertson (2010) for the active performance of governance.

11. T. Simpson, personal interview, August 2, 2011, Santa Barbara, NM. This aligns with Correia (2013) and Blomley (2013) on the performative aspects of property. Property is not just an object. Perceptions of property change depending on the angle of perception, like a prism, as is the case for water in New Mexico (Merrill 2011).

12. Rodriquez is the author of *Acequia* (2006), which documented the Taos Valley acequias and their struggle to maintain water in ditches.

13. E. Avaro, personal interview, August 13, 2011, Alcalde, NM.

14. At least eleven parciantes shared this opinion as to why their ditch had not participated in the regional organization scheme.

15. New Mexico Statutes Annotated (NMSA) 73-2-47 (1915).

16. Section 10-15-1(B) of the Open Meetings Act (NMSA 1978, sections 10-15-1 to 10-15-4).

17. P. Trujillo, personal interview, August 2012, near Jemez River.

18. This total count (thirty-eight) is based on close textual analysis of the interviews performed by the author.

19. D. Benavides, attorney at NM Legal Aid, July 28, 2015, Taos, NM.

20. D. Benavides, attorney at NM Legal Aid, February 3, 2010, Santa Fe, NM.

21. In the 2003 state water plan, for example: "For the 21st century, the State must develop water market and water banking mechanisms that will facilitate the voluntary

movement of water from old uses to new, with the marketplace supplying the appropriate rewards and the State providing the necessary safeguards." This can be found in the preliminary materials found on page 2 of the New Mexico State Water Plan, December 3, 2003, www.ose.state.nm.us/Planning/SWP/PDF/2003StateWaterPlan.pdf.

22. L. Drury, attorney for developers, during testimony before the New Mexico state subcommittee on natural resources and water, July 28, 2015, Taos, NM.

23. Connie Ode, attorney for acequias, during testimony before the New Mexico state subcommittee on natural resources and water, July 28, 2015, Taos, NM.

24. Elliot (1943).

25. See Peña (1999), McCarthy (2005), and Bakker (2007) for more on how commons and common-use practitioners resist the individual "property rights" demands of capitalism.

26. Water continues to be locally constructed and defended as a commons. Thus, the water commons can survive not just as an object of use but as a central location of relationships in preserving communal links (following McCarthy 2005; Turner 2016).

27. NMSA (2008), 73-2-55.1. Water banking measures (enacted in 2003).

28. Forfeiture cases brought by the state of New Mexico are rare. Yet the state code empowers the OSE to declare water rights as forfeit if unused for more than five years.

29. Now included in state statutes 72-1-43 and 72-14-44 N.M.S.A., Cum Supp. 1993.

30. See Molle (2009) for more on the challenges of watershed planning; see, by way of comparison, Powell's visions of watershed democracies in the American West (Worster 2003).

31. These are my perceptions after having attended the statewide meeting and several regional planning meetings and recording statements that captured these sentiments.

32. See Chief et al. (2016) for more on how engaging productively with sovereign tribal nations in the Southwest will be critical in this century.

33. Raul Arellano, interview, December 5, 2010, Taos, NM.

34. J. A. Morea, personal interview, July 28, 2016, Santa Fe, NM.

35. L. Dorcet, personal communication, August 5, 2014, Santa Fe, NM.

36. T. Maduro, personal communication, July 24, 2011, Embudo, NM.

37. A. Pino, personal interview, February 2, 2010, Las Vegas, NM. See also Liverman (2004) for concerns when resources come under state or market-driven governance regimes.

38. D. Salinas, personal communication, August 15, 2012, Pojoaque Valley, NM.

39. As reported by Kay Matthews, "Acequia Parciantes Protest at Abeyta Implementation Meeting over Mitigation and ASR Wells," *La Jicarita,* April 28, 2017, https://lajicarita.wordpress.com/2017/04/28/acequia-parciantes-protest-at-abeyta-implementation-meeting-over-mitigation-and-asr-wells/.

40. Rodriguez (2006). The new presence of Hispano and Anglo-American college degree holders has allowed the acequias to adapt to the modernizing expectations of the state of New Mexico. Tellingly, these are both trends that Meinzen-Dick (2007) found in India: social capital among a single religion was a strong basis for cooperation, but an increase in college education was effective at organizing water user organizations.

41. Residents along a Taos Valley acequia are concerned about a water transfer from the village of Questa to aquifer recovery wells on the outskirts of Taos. Members of the acequia in question openly worry about the annual costs of these recovery wells. See "Taos Acequia

Still Objects to Terms of Abeyta Settlement," *La Jicareta,* https://lajicarita.wordpress
.com/2016/07/14/taos-acequia-still-objects-to-terms-of-abeyta-settlement/.

42. At least eighty-five interviewees mentioned these groups as unintended benefits or products of the lingering dissatisfaction with water adjudication in the state.

CHAPTER 8. CITY WATER, NATIVE WATER, AND THE UNKNOWN FUTURE

1. T. Roybal, personal interview, August 14, 2014, Santa Fe, NM.

2. See Bowden (1977) and Glennon (2002) for more on groundwater "mining" in the Greater Southwest.

3. Historian Daniel Tyler (1990) has critiqued the assumed Spanish policy or law that remained quietly implicit yet functional for decades as cities in New Mexico were able to keep acquiring a greater share of water during the twentieth century. For a wider-ranging treatment of Spanish and Mexican water policies and approaches, see Meyer (1984).

4. California did recognize municipal higher rights to water.

5. A good summary of the overturned pueblo rights water doctrine is provided by Brockmann and Martinez in Ortega Klett's (2012, 109–122) edited volume. See also Mulvaney (2005) for the long-standing concerns of the city of Las Vegas in this overturning. Oddly, the pueblo rights doctrine for cities is still followed in the state of California.

6. For an excellent summary of the Martinez case; the use of it by Las Vegas, New Mexico; and the demise of the imaginary pueblo rights doctrine, see Mulvaney (2005).

7. A 1955 technical report by Wells Hutchins (1955, 7–8), commissioned by the OSE, indicated that Santa Fe had no basis and made no claims to a municipal "pueblo rights doctrine." Las Vegas' claim to the pueblo water rights doctrine went largely unchecked as it developed between the 1958 Cartright decision and the eventual defeat of the doctrine in 2004 (Martinez). See Mulvaney (2005) for more.

8. See Snow (1988). A similar pattern happened on the Pecos River. As a farmer from the upper Pecos told Em Hall (2002, 249): "Irrigated agriculture has been a way of life up here for more than two centuries. It dropped off a lot after the Second World War when the local subsistence economies collapsed."

9. For more, see Romero Pike's (2010, 42) short essay on the deprivation of ditch water by the city of Santa Fe.

10. D. Roybal, personal interview, August 2, 2011, Santa Fe, NM.

11. Tomás is only distantly related to the David Roybal mentioned previously.

12. T. Roybal, personal interview, August 14, 2014, Santa Fe, NM.

13. B. Gallegos, personal interview, August 4, 2012, Santa Fe, NM.

14. V. Gurulé, personal interview, January 20, 2010, Santa Fe, NM.

15. C. Ramsey, personal interview, August 3, 2013, Santa Fe, NM.

16. T. Santistevan, interview with the author, October 16, 2010, Santa Fe, NM.

17. Pending and proposed legislation, State of New Mexico Senate Memorial 70, 1st sess., 2017, found at https://www.nmlegis.gov/Sessions/17%20Regular/memorials/senate /SM070.html.

18. The Middle Rio Grande Conservancy District (MRGCD) has an agreement with nearby Pueblos for prior and paramount irrigation rights on certain tracts of Indian land. Overall, Indian water rights have not been quantified as of 2017.

19. Romero-Wirth and Kelly (2012., 13). The original document, part of the Utton Law Center's *WaterMatters!* publication, can be found at "Water Rights Management in New Mexico and along the Middle Rio Grande: Is AWRM Sufficient?," http://uttoncenter.unm .edu/pdfs/water_rights_mgmt.pdf.

20. See Clark's (1987) magnum opus, especially chapter 12 (188–213), for a comprehensive discussion on the MRGCD in the context of flooding in the 1930s.

21. These federal water rights are recognized by the MRGCD and are separate from any future state-driven adjudication that may award further water rights.

22. The MRGCD, and what it did to traditional ditches and farmers, was mentioned by sixteen interviewees in Taos.

23. The MRGCD presents its own version of history, asserting on its web "history" page that "the existing irrigation systems [acequias] were insufficient and primitive" at the time the MRGCD was formed.

24. A clear description of this process of incorporation is offered by Sam Markwell in "Water Users Suing Middle Rio Grande Conservancy District: The Devil Is in the Details," *La Jicarita,* October 18, 2012, https://lajicarita.wordpress.com/2012/10/18/abq-south-valley -farmers-suing-middle-rio-grande-conservancy-district-the-devil-is-in-the-details/.

25. M. Santiz, former water division employee for Albuquerque, personal communication, August 15, 2011, Albuquerque, NM.

26. The other challenging aspect is that some twenty-one thousand acre-feet of water rights have been sold and transferred to Albuquerque and Santa Fe over a thirty-year period (1982–2011, as data were available). See the document prepared by the Utton Center for the difficulties of a true water market in New Mexico: http://uttoncenter.unm.edu/pdfs/water -matters-2015/16_Water_Marketing.pdf.

27. Flood control is a typical goal for many dams, yet the argument that these large projects were for development purposes comes directly from Kupel (2003). But Kupel's arguments, along with those about Denver's push for urban water in Limerick and Hanson's (2012), run counter to the notion that these were hydraulic empires of bureaucracy serving only capitalism (following Worster 1985). Both Kupel and Limerick and Hanson argue that local forces were vital to pushing larger state and federal infrastructure efforts.

28. See Bakker (2004, 2007, 2010) for more on the differences and distinctions between pricing, commoditization, and the privatization of water.

29. L. Diane, personal interview, August 15, 2014, Albuquerque, NM.

30. As of April 2017, see more at "Santolino Master Plan," Bernalillo County Planning and Development Services, www.bernco.gov/planning/proposed-santolina-level-a-master -plan.aspx.

31. Anonymous. This was stated publicly at the January 2017 meetings of the New Mexico Water Dialogue in Albuquerque.

32. The Yavapai successfully opposed the never-built Orme Dam in Arizona, discussed in Espeland (1998).

33. Pecos (2007). For more details on Cochiti, see Ebright et al. (2014) and Phillips et al. (2011).

34. Anonymous, 2010. This person did not want even a pseudonym attached to this quip made in passing.

35. Gerald Toloaque, personal interview, January 17, 2010, Cochiti Pueblo, NM.

36. Phillips et al. (2012).

37. See Pecos (2007, 650–651) for more on the remarkable archival document recovered during Aamodt research that led to a spring being restored to Cochiti Pueblo.

38. Hall (2002, 217) once wrote: "Compared to the preemptive assertion of federal water rights on the Rio Grande, the Pecos River is a model, temporarily at least, of intergovernmental civility."

39. R. Snowe, personal interview, August 12, 2011, Truth or Consequences, NM.

40. C. Travers, personal communication via email, July 18, 2016.

41. See former Judge Jerald Valentine's direct piece (chap. 2, 29–51) in the larger edited volume by Ortega Klett (2012).

42. See Phillips et al. (2011) for the Rio Grande and the earlier Hall (2002) for these respective interstate challenges between New Mexico and Texas.

43. Estimates based on the 2013 OSE report *2010 New Mexico Water Use by Categories,* Technical Report 54, Office of the State Engineer, Santa Fe, NM.

44. Among others, Michael Pease (2010) and David Shiveley (2001) have tracked such active water markets along the Middle Rio Grande. Brookshire et al. (2013) also discuss markets and policy for the Rio Grande.

45. See Erik Swyngedouw's (2004, 2005, 2009) work on the urbanization of water, in which capitalism has deep stakes in moving water from the countryside to the city in almost all countries.

CHAPTER 9. BEYOND ADJUDICATION: NATURE'S SHARE OF WATER

1. W. Benavides, personal interview, November 11, 2009, Las Vegas, NM.

2. See deBuys (2008) for an overview of the Anthropocene in the Southwest. For the origins of the term, see Crutzen (2002).

3. See deBuys (2008, 2011) for more information on the Anthropocene.

4. For a good overview of these concerns, see Dettinger et al. (2015). Water managers are thus already mobilized to think about scarcity, or "drought," as a concern (see Stroup 2011 for the latter).

5. Ingram and Malamud-Roam (2013).

6. Ibid.

7. IPCC (2013).

8. See *Santa Fe Basin Study: Adaptations to Projected Changes in Water Supply and Demand,* US Department of the Interior Bureau of Reclamation, 2015, https://www.usbr .gov/watersmart/bsp/docs/finalreport/SantaFe/Santa-Fe-Basin-Final.pdf.

9. See Stroup (2011) for more.

10. See Bureau of Reclamation (2005). It used to be publicly available. This report has since been removed from the Department of Interior web site.

11. Rockstrom et al. (2014) have expressed these concerns as trying to plan some resiliency into the "water drama" of our current era.

12. See McCarthy (2005) for an argument on how the commons are one of the last sites of resistance against increased resource commodification. See also the more recent review

by Turner (2016) on the actual acts of "commoning" that are attached to these strategies of resisting parsing property to individuals.

13. E. Gomez, personal interview, November 11, 2009, Las Vegas, NM.

14. See also Birkenholtz (2008, 2009) for more on contested expertise. Earlier work by Robbins (2000) revealed the local/state expertise binary is rarely clear-cut.

15. Kupel's (2003) phrasing of water as "fuel for growth" comes from his book about urbanization in Arizona. The same case about how water has been viewed as merely a catalyst for further urban development holds for Albuquerque. See Price (2014) for several examples.

16. See Cosens et al. (2014) for one framing that calls for more transparency in creating interdisciplinary dialogue and collaboration. Baird et al. (2015) call for more pragmatism when it comes to managing and governing water. See also Turner's (2014) recent discussion about the possible opportunities for political ecologists and resilience scholars to collaborate on land-use ecology situations, even if resilience theorists tend to emphasize "systems thinking" more than political ecologists typically do (see Rockstrom et al. 2014).

17. See, for example, Simon (2017), which starts with the 1991 Tunnel fire in Oakland and treats the last twenty-five years of wildland urban interface fires.

18. See Graf (1994) for more on the naturally high plutonium concentrations along parts of the Rio Grande. Lake Cochiti itself is now a legacy contaminant pool of sediment high in radioactive particles.

19. See Hall (2002) for more on the Pecos River struggle between the states.

20. The attorney wanted absolute anonymity, not just a pseudonym.

21. In 1998, the state attorney general for New Mexico lodged an opinion that state statute does allow the state engineer to consider instream flows a beneficial use; this opinion was crucial to managing the silvery minnow in the Rio Grande as an endangered species. Most western states face challenges in balancing the federally mandated needs for endangered species (see Tarlock 1985). This yet again reflects the legacy of river federalism in the United States (Doyle et al. 2013).

22. As argued and illustrated in Kosek (2006).

23. See Groenfeldt (2010) regarding water ethics as a way to address ecological governance in a pragmatic way, illustrated in Santa Fe's living river approach. Glennon (2006) describes the Endangered Species Act and the challenges it offers for managing water. On a more abstract level, see Schmidt's (2017) work on how liberal constructions of water as a resource produced this dilemma in the first place.

24. F. Schwartz, personal interview, July 12, 2012, Albuquerque, NM.

25. A good summary of the minnow wars can be found in DuMars's chapter in Ortega Klett (2012).

26. Paul Tashjian, former United States Fish and Wildlife Service biologist, email communication, January 2017.

27. See the recent Stanford Woods Institute for the Environment report *Water in the West*, Colorado River Basin Environmental Water Transfers Scorecard, March 2017, http://waterin thewest.stanford.edu/sites/default/files/Co_River_Basin_Env_Transfers_Scorecard.pdf.

28. B. Anderson, personal communication, August 13, 2011, Albuquerque, NM.

29. See Stroup (2011) for more on how water managers in different parts of the United States plan to manage water under drought or climate-change scenarios.

30. See Plewa (2009, chap. 8) for more on the living river program in Santa Fe before it was actualized as a program. Her analysis previews what was to come just a few years later. These 1000 AFY are subject to curtailment in times of water shortages.

31. A. Oglisi-Danforth, personal communication, July 2, 2014, Santa Fe, NM.

32. Moments of drought or water "crises" are opportunities to rethink governance, technology, and social institutions—a point made most recently by Fleck (2016), along with Taylor and Sonnenfeld (2017).

33. This living river program was largely the result of citizen pressure through the Santa Fe River Watershed Association. This is part of a national trend in river restoration (see McCool 2012).

34. Gleick (2010) summarizes what a sustainable plan for water management in the Southwest would look like. See Christian-Smith et al. (2012) on arguing for a national water policy.

35. For more on these groups, see the Freshwater Trust, www.thefreshwatertrust.org/; Colorado Water Trust, http://www.coloradowatertrust.org; and Raise the River, http://raisetheriver.org/. See Fleck (2016) for a rather optimistic reinterpretation of the Colorado River situation. See Ward's (2015) reprinted book for the long and sad history of the death of the Colorado River Delta.

36. See Rebecca Lave's (2012) treatment of privatization in watershed restoration.

37. A. Velarde, personal interview, August 15, 2011, near San Lorenzo (Mimbres), NM.

38. Ingram and Brown (1987).

39. L. Indres, personal interview, August 8, 2014, near Embudo, NM.

40. T. Gaviola, personal interview, August 12, 2011, Silver City, NM.

41. A. Velez, personal interview, January 8, 2010, near Las Vegas, NM. This was not a rare observation; at least twenty-two other irrigators made similar comments.

42. See Margolis et al. (2011) and Woodhouse et al. (2013) for examples of how dendrochronology (tree ring data) can inform water managers about past climate-water quantity relationships in New Mexico. What stands out from most of these data is that the twentieth century, when we built water law and water infrastructure, was remarkably wet compared to centuries past.

43. Laura Paskus produced a series of entries on New Mexico as the vanguard of climate-change effects on water, and these can currently be found at New Mexico In Depth, http://nmindepth.com/2015/10/14/new-mexico-is-at-the-cutting-edge-of-change-climate-change/.

44. See "Water Wiser: New Smart Water Meters Aimed at Spotting Leaks before Bills Climb," *Santa Fe Reporter*, www.sfreporter.com/santafe/article-10584-water-wiser.html, for a recent example about improving urban metering devices for municipal supplies. Some recent interdisciplinary work by Hess et al. (2016), however, offers some sobering analysis that supply-side solutions of finding water elsewhere continues to lure cities and utilities.

45. A. Graham, personal interview, August 9, 2011, Santa Fe, NM.

46. This inability for any state, or State, to control water within a territorial bounded envelope is referred to by Mason and Khawlie (2016) as *fluid sovereignty*; water cannot, by its nature, be contained.

47. Grantham and Viers (2014) approximated that the state of California has allocated paper water rights to five times the actual supply of available wet water.

48. For more on this, see "The State Has Issued 7,910 Cutoff Notices to Owners of Water Rights in Watersheds across the State," *Willits News,* July 18, 2014, www.willitsnews.com /drought/ci_26173104/state-has-issued-7-910-cutoff-notices-owners. Echeverria (2014) has also raised concerns that claims of governmental "takings" may rise if water shortages induced by climate change lead to curtailment. If the state engineer has to reduce water rights, will irrigators, industries, and cities sue? Or will a new mechanism emerge that reduces these takings claims? This is similar to the concerns in the Pojoaque related to the Aamodt settlement terms to cap post-1982 groundwater wells.

49. This seems humorous, but the intent of keeping evaporation down and keeping pests out for water quality purposes is sound. See "Why Did L.A. Drop 96 Million 'Shade Balls' into Its Water?," *National Geographic,* August 12, 2015, http://news.nationalgeographic .com/2015/08/150812-shade-balls-los-angeles-California-drought-water-environment/.

50. See Tidwell et al. (2014) for a recent, useful modeling study on western water availability that draws on large data sets.

51. California has changed its regulatory stance on surface and groundwater, to its credit. Further technifying water solutions will only beget more engineering, rather than provide a real solution to the amount of water actually available. For a wider comparison of several states on groundwater, see also Sugg et al.'s (2016) recent work.

52. See MacDonnell's (2015) wide-ranging piece on adjudications; also Tarlock (2010) on climate-related issues.

53. W. Benavides, personal interview, November 11, 2009, Las Vegas, NM.

CHAPTER 10. WATER CODA, WITH NO END IN SIGHT

1. Lucy Moore (2013, 185).

2. See Merchant (2016) for more on the continued challenges of autonomous nature. See also McPhee (1989) for this illusory "control of nature" that has no real permanency to it.

3. A point noted also by Phillips et al. (2011). Fleck (2016) has similar thoughts about the Colorado River.

4. A point repeatedly made by Bryan (2015) in her review of western adjudication procedures; the data produced are often obsolete, incomplete, and may need to be updated in most states.

5. J. Fresquez, personal interview, October 11, 2009, Taos, NM.

6. See Robison et al. (2018, forthcoming) for more on the concept of "water colonialism" in the West and in the international context.

7. MacKenzie (2006) argued that economic models serve as drivers for economic speculation and change, not just as "snapshot" (camera) understanding of a national economy. Adjudication likewise serves as an engine of change as much (or more) as it serves as the state's snapshot understanding of water use and water citizens in New Mexico.

8. As informed by legal approaches in anthropology (Mattei and Nader), geography (Blomley, Delaney), and science studies (Latour).

9. See O'Neill (2006) for more on the federal role in flood protection and crafting rivers into territorial spaces of state power.

10. See White (1995) for the "organic machine" metaphor of transforming the western rivers, in his case the Columbia River.

11. Yet see Schorr (2012) for more on the equitable origins of prior appropriation. The system was created to avoid water hoarding and monopolies.

12. Montoya (2002, 217).

13. Or as Wilder and Lankao (2006, 1977) summarized institutional changes to water in Mexico: "The creation of new forms of water institutions requires not the retrenchment of the state but rather its involvement to ensure accountability, transparency, equity, and sustainability." Part of this engagement will have to come in the form of information sharing.

14. See Gleick (2010) for a still-relevant account of how we might manageably supply the Southwest with new, reused, and treated water in creative ways. For a longer technical treatise that fully considers the costs and politics of water, see also Sedlak (2014).

15. I agree with Agnew (2011, 474) in that we need "a practical conception of politics as both a focus for analyzing the actual ways in which water provision is subject to dispute and as a normative commitment to actively shaping the world through popular participation."

16. Phillips et al. (2011) made similar observations about the limited capacity for technology to "fix past and current problems" as related to water resources. Where my view differs is that institutions, objects, *and* law are all forms of technology. Administrative law, including adjudication, and settlement further encourage state and federal agencies to add infrastructure (see Beard 2015 for a vitriolic critique).

17. As noted by Hess et al. (2016), this addiction to producing "new water" is merely a transfer of water from somewhere else and not actual new net water.

18. Strang (2009), using ethnography, similarly found that "agency and identity" remain central to water users in Australia.

19. B. Miera, personal interview, October 27, 2009, Santa Fe, NM.

20. Rodriguez (1990).

21. From Utton Center (2013); one of dozens of estimates regarding the remaining length of adjudications before the state has complete knowledge of water and water users in New Mexico.

22. Some ninety-three interviewees cited some version of this saying.

23. See Bryan (2015, 510) for more details on how the calculation of consumptive use of water by New Mexico's agency may streamline later changes in water use or changes to the location of use.

24. A state representative for New Mexico stated this at the January 11, 2018, meeting of the New Mexico Water Dialogue group in Albuquerque, New Mexico.

25. See *E&E News*, "Water Rights: Special Master Urges Supreme Court to Weigh Texas-N.M. Dispute," February 10, 2017, http://www.eenews.net/stories/1060049896 for more on recent developments between New Mexico and Texas in their struggles over the Lower Rio Grande and water stored in Elephant Butte Dam. The full report to the US Supreme Court is authored by Gregory Grimsal, "On New Mexico's Motion to Dismiss Texas's Complaint and the United States' Complaint in Intervention and Motions of Elephant Butte Irrigation District and El Paso County Water Improvement District No. 1 for Leave to Intervene," *First Interim Report of the Special Master*, no. 141, February 9, 2017.

26. See Fleck (2016) for more on this 2015 dam release that temporarily sent water all the way down the Colorado River to the Sea of Cortez.

27. As stated by Senator Ron Griggs, July 27, 2015, Taos, New Mexico.

28. See Conca (2015, 314); as he puts it: "In both the institutions of administrative water governance and the institutionalised political economy of financing water infrastructure, a pre-greenhouse logic prevails. In the former instance, the governance system has been moving toward integrated mechanisms of increasing complexity and interconnectedness; in the latter, the financing system has been moving toward volatile private capital markets." See Barnhart et al. (2016) for the tight coupling between snowpack and surface stream flows.

29. See the work on groundwater by Emel and various colleagues (1988, 1992, 1995) and more recent works by A. Wolf (2000), Norman and Bakker (2009), and Wilder et al. (2010), which speak to international implications for water management.

30. B. Tafoyeta, personal communication, NM Water Dialogue, January 2011, Albuquerque, NM. See Kate Berry's (2012) survey of the evolving relationship between tribes and water and their newfound ability to enforce sovereign water-quality benchmarks on shared rivers.

31. As Hughes (2017, 261) put it regarding the advantage of water settlements over adjudications, "Of course settlements in cases such as Aamodt and Abeyta have included substantial federal funding for the Pueblos to be able to develop the rights confirmed to them, something that an adjudication would not have afforded them."

32. See Barnes (2014) on these points for Egypt's Nile water uses.

33. J. Encinitas, personal interview, October 31, 2009, Mora Valley, NM.

34. See Ortiz et al. (2007) for more on rural acequia agricultural decline.

35. See Fernald et al. (2007, 2012) for examples of the hydrologic connectivity between acequias to groundwater.

36. This aspect of a nonliberal commons, one not deriving solely from "private" benefits or individuals, has been argued for by Schmidt and Mitchell (2014). The point here is that acequias were already doing this as a form of practice and not just in theory.

37. To echo Levine (2008, 261), "The local perspective of the acequias has a contribution to make to regional, national, and international issues. Water-sharing agreements recorded in northern New Mexico legal cases are similar to agreements found throughout the world. Increasingly public-policy administrators are turning to indigenous peoples and local perspectives in drafting resource-management agendas."

38. See Crossland (1990) for examples of this. This logic is alive throughout Latin America; see Perreault (2005, 2008) and Prieto (2016) for examples from Peru and Chile, respectively, on how customary practice and law can stay embedded in ditch communities.

39. See Bartel (2016) for a review of critical legal geography; adjudication practices fit neatly into administrative law practices yet remain underexamined in scholarly literature outside of law school journals.

REFERENCES

Agnew, J. 2011. Waterpower: Politics and the Geography of Water Provision. *Annals of the Association of American Geographers* 101(3): 463–476.

Agrawal, A. 2005. *Environmentality: Technologies of Government and the Making of Subjects.* Durham, NC: Duke University Press.

Aguilar-Robledo, M. 2009. Contested Terrain: The Rise and Decline of Surveying in New Spain, 1500–1800. *Journal of Latin American Geography* 8(2): 23–47.

Akhter, M. 2016. Desiring the Data State. *Transactions of the Institute of British Geographers* 42:1–13. doi:10.1111/tran.12169.

Alatout, S., 2007. State-ing Natural Resources through Law: The Codification and Articulation of Water Scarcity and Citizenship in Israel. *Arab World Geographer* 10:16–37.

———. 2008. "States" of Scarcity: Water, Space, and Identity Politics in Israel, 1948–59. *Environment and Planning D: Society and Space* 26:959–982.

Archambault, S., and Ulibarri, J. 2007. Nonmarket Valuation of Acequias: Stakeholder Analysis. *Environmental Engineering and Management Journal* 6(6): 491–495.

Arellano, J. E. 2014. *Enduring Acequias: Wisdom of the Land, Knowledge of the Water.* Albuquerque: University of New Mexico Press.

Baird, J., Plummer, R., Bullock, R., Dupont, D., Heinmiller, T., Jollineau, M., Kubick, W., Renzetti, S., and Vasseur, L. 2016. Contemporary Water Governance: Navigating Crisis Response and Institutional Constraints through Pragmatism. *Water* 8(224): 1–16.

Bakker, K. 2004. *An Uncooperative Commodity: Privatizing Water in England and Wales.* Oxford: Oxford University Press.

———. 2007. The "Commons" versus the "Commodity": Alter-globalization, Anti-privatization and the Human Right to Water in the Global South. *Antipode* 39(3): 430–455.

———. 2010. *Privatizing Water: Governance Failure and the World's Urban Water Crisis.* Cornell, NY: Cornell University Press.

Banister, J. 2011. Deluges of Grandeur: Water, Territory, and Power on Northwest Mexico's Río Mayo, 1880–1910. *Water Alternatives* 4(1): 35–53.

———. 2014. Are You Wittfogel or Against Him? Geophilosophy, Hydro-sociality, and the State. *Geoforum* 57:205–214.

Barker, J. 2011. *Native Acts: Law, Recognition, and Cultural Authenticity*. Durham, NC: Duke University Press.

Barnes, J. 2014. *Cultivating the Nile: The Everyday Politics of Water in Egypt*. Durham, NC: Duke University Press.

Barnhart, T. B., Molotch, N. P., Livneh, B., Harpold, A. A., Knowles, J. F., and Schneider, D. 2016. Snowmelt Rate Dictates Streamflow. *Geophysical Research Letters* 43:1–11.

Barrett, E. M. 2002. *Conquest and Catastrophe: Changing Rio Grande Pueblo Settlement Patterns in the Sixteenth and Seventeenth Centuries*. Albuquerque, NM: University of New Mexico Press.

Bartel, R. 2016. Legal Geography, Geography, and the Research-Policy Nexus. *Geographical Research* 54(3): 233–244.

Bauer, C. 2004. *Siren Song: Chilean Water Law as a Model for International Reform*. Washington, DC: Resources for the Future Press.

———. 2010. Market Approaches to Water Allocation: Lessons from Latin America. *Journal of Contemporary Water Research and Education* 144:44–49.

Baxter, J. 1997. *Dividing the Waters: 1700–1912*. Albuquerque: University of New Mexico Press.

———. 2000. Measuring New Mexico's Irrigation Water: How Big is a "Surco"? *New Mexico Historical Review* 75(3). 397–413.

Beard, D. P. 2015. *Deadbeat Dams: Why We Should Abolish the U.S. Bureau of Reclamation and Tear Down Glen Canyon Dam*. Boulder: Johnson Books.

Benda-Beckmann, F. v., v. Benda-Beckmann K., and Griffiths, A. M. O., eds. 2009. *Spatializing Law: An Anthropological Geography of Law in Society*. Surrey/Burlington: Ashgate.

Benson, R. 2012. Alive but Irrelevant: The Prior Appropriation Doctrine in Today's Western Water Law. *University of Colorado Law Review* 83(3): 675–714.

Berry, K. A. 2012. Tribes and Water. In *A Twenty-First Century U.S. Water Policy*, eds. J. Christian-Smith, P. H. Gleick, H. Cooley, L. Allen, A. Vanderwarker, and K. A. Berry, pp. 90–108. New York: Oxford University Press.

Birkenholtz, T. 2008. Contesting Expertise: The Politics of Environmental Knowledge in Northern Indian Groundwater Practices. *Geoforum* 39(1): 466–482.

———. 2009. Irrigated Landscapes, Produced Scarcity, and Adaptive Social Institutions in Rajasthan, India. *Annals of the Association of American Geographers* 99(1): 118–137.

Blair, J. A. 2017. Settler Indigeneity and the Eradication of the Non-Native: Self-Determination and Biosecurity in the Falkland Islands (Malvinas). *Journal of the Royal Anthropological Institute* 23(3): 580–602.

Blomley, N. 2003. Law, Property, and the Geography of Violence: The Frontier, the Survey, and the Grid. *Annals of the Association of American Geographers* 93(1): 121–141.

———. 2008. Simplification Is Complicated: Property, Nature, and the Rivers of Law. *Environment and Planning A* 40(8): 1825–1842.

———. 2013. Performing Property: Making the World. *Canadian Journal of Law and Jurisprudence* 26(1): 23–48.

Blomley, N., Delaney D., and Ford, R., eds. 2001. *The Legal Geographies Reader: Law, Power and Space*. Oxford: Blackwell.

Blomquist, W., and Schlager, E. 2005. Political Pitfalls of Integrated Watershed Management. *Society and Natural Resources* 18(2): 101–117.

Boelens, R. 2009. The Politics of Disciplining Water Rights. *Development and Change* 40(2): 307–331.

Boelens, R., Getches, D., and Guevara-Gil, A., eds. 2010. *Out of the Mainstream: Water Rights, Politics and Identity*. London: Earthscan.

Bowden, C. 1977. *Killing the Hidden Waters*. Austin: University of Texas Press.

Brewer, J., Glennon, R., Ker, A., and Libecap, G. 2008. Water Markets in the West: Prices, Trading, and Contractual Forms. *Economic Inquiry* 46(2): 91–112.

Brienza, S. D. 1992. Wet Water vs. Paper Rights: Indian and Non-Indian Negotiated Settlements and Their Effects. *Stanford Environmental Law Journal* 11:151–199.

Brockman, J. C., and Martinez, E. 2012. Las Vegas, New Mexico: The Rise and Fall of the Pueblo Water Rights Doctrine. In *One Hundred Years of Water Wars in New Mexico, 1912–2012*, ed. C. T. Ortega Klett, pp. 109–122. Santa Fe: Sunstone Press.

Brooks, J. F. 2002. *Captives and Cousins: Slavery, Kinship, and Community in the Southwest Borderlands*. Chapel Hill: University of North Carolina Press.

Brookshire, D., Gupta, H., and Matthews, O. P. 2013. *Water Policy in New Mexico: Addressing the Challenge of an Uncertain Future*. London: Routledge Press.

Brown, C. J., and Purcell, M. 2005. There's Nothing Inherent about Scale: Political Ecology, the Local Trap, and the Politics of Development in the Brazilian Amazon. *Geoforum* 36(5): 607–624.

Brown, F. L., and Ingram, H. 1987. *Water and Poverty in the Southwest*. Tucson: University of Arizona Press.

Bryan, M. 2015. At the End of the Day: Are the West's General Stream Adjudications Relevant to Modern Water Rights Administration? *Wyoming Law Review* 15(2): 461–516.

Budds, J. 2009. Contested H2O: Science, Policy and Politics in Water Resources Management in Chile. *Geoforum* 40(3): 418–430.

———. 2013. Water, Power and the Production of Neoliberalism in Chile, 1973–2005. *Environment and Planning D: Society and Space* 31:301–318.

Bureau of Reclamation. 2005. *Water 2025*. Washington, DC: US Department of the Interior.

Carroll, P. 2012. Water and Technoscientific State Formation in California. *Social Studies of Science* 42(4): 489–516.

———. 2006. *Science, Culture, and Modern State Formation*. Berkeley: University of California Press.

Carter, E. D. 2008. State Visions, Landscape, and Disease: Discovering Malaria in Argentina, 1890–1920. *Geoforum* 39(1): 278–293.

———. 2012. *Enemy in the Blood: Malaria, Environment, and Development in Argentina*. Tuscaloosa: University of Alabama Press.

Chief, K., Meadow A., and Whyte, K. 2016. Engaging Southwestern Tribes in Sustainable Water Resources Topics and Management. *Water* 8(8): 350.

Christian-Smith, J., Gleick, P., Cooley, H., Allen, L., Vanderwarker, A., and Berry, K. A. 2012. *A Twenty-First Century U.S. Water Policy*. Oxford: Oxford University Press.

Clark, I. 1987. *Water in New Mexico: A History of Its Management and Use.* Albuquerque: University of New Mexico Press.

Clarke, J. 2010. Enrolling Ordinary People: Governmental Strategies and the Avoidance of Politics? *Citizenship Studies* 14(6): 637–650.

Cohen, A., and Davidson, S. 2010. The Watershed Approach: Challenges, Antecedents, and the Transition from Technical Tool to Governance Unit. *Water Alternatives* 4(1): 1–14.

Colby, B. G., McGinnis, M. A., and Rait, K. 1989. Procedural Aspects of State Water Law: Transferring Water Rights in Western States. *Arizona Law Review* 31(4): 697–720.

Colby, B. G., Thorson, J. E., and Britton, S., eds. 2005. *Negotiating Tribal Water Rights: Fulfilling Promises in the Arid West.* Tucson: University of Arizona Press.

Conca, K. 2006. *Governing Water: Contentious Transnational Politics and Global Institution Building.* Cambridge, MA: MIT Press.

———. 2015. Which Risks get Managed? Addressing Climate Effects in the Context of Evolving Water-Governance Institutions. *Water Alternatives* 8(3): 301–316.

Correia, D. 2013. *Properties of Violence: Law and Land Grant Struggle in Northern New Mexico.* Athens: University of Georgia Press.

Cosens, B., Gunderson L., Allen C., and Benson, M. H. 2014. Identifying Legal, Ecological and Governance Obstacles, and Opportunities for Adapting to Climate Change. *Sustainability* 6:2338–2356.

Coulthard, G. S. 2014. *Red Skin, White Masks: Rejecting the Colonial Politics of Recognition.* Minneapolis: University of Minnesota Press.

Cox, M. 2014. Modern Disturbances to a Long-Lasting Community-Based Resource Management System: The Taos Valley Acequias. *Global Environmental Change* 24:213–222.

Craib, R. 2004. *Cartographic Mexico: A History of State Fixations and Fugitive Landscapes.* Durham, NC: Duke University Press.

Crawford, S. 1988. *Mayordomo: Chronicle of an Acequia in Northern New Mexico.* Albuquerque: University of New Mexico Press.

———. 1990. Dancing for Water. *Journal of the Southwest* 32(3): 265–267.

———. 2003. *The River in Winter: New and Selected Essays.* Albuquerque: University of New Mexico Press.

Crossland, C. 1990. Acequia Rights in Law and Tradition. *Journal of the Southwest* 32(3): 278–287.

Crutzen, P. J. 2002. Geology of Mankind. *Nature* 415:23–23.

Davis, D. 2009. Historical Political Ecology: On the Importance of Looking Back to Move Forward. *Geoforum* 40:285–286.

Dawson, A., Zanotti, L., and Vaccaro, I., eds. (2014). *Negotiating Territoriality: Spatial Dialogues between State and Tradition.* Routledge Studies in Anthropology. London: Routledge Press.

deBuys, W. 2008. Welcome to the Anthropocene. *Rangelands* 30(5): 31–35.

———. 2011. *A Great Aridness: Climate Change and the Future of the American Southwest.* New York: Oxford University Press.

DeJong, D. H. 2016. *Stealing the Gila: The Pima Agricultural Economy and Water Deprivation, 1848–1921.* Tucson: University of Arizona Press.

Delaney, D. 2003. *Law and Nature.* Cambridge: Cambridge University Press.

————. 2010. *The Spatial, the Legal and the Pragmatics of World-Making*. New York: Routledge Press.

Dettinger, M., Udall, B., and Georgakakos, A. 2015. Western Water and Climate Change. *Ecological Applications* 25(8): 2069–2093.

Doolittle W. E. 2000. *Cultivated Landscapes of Native North America*. New York: Oxford University Press.

Doyle, M., R. Lave, M. Robertson, and Ferguson, J. 2013. River Federalism. *Annals of the Association of American Geographers* 103(2): 290–298.

DuMars, C., O'Leary, M., and Utton, A. E. 1984. *Pueblo Indian Water Rights: Struggle for a Precious Resource*. Tucson: University of Arizona Press.

Dunbar-Ortiz, R. 2007. *Roots of Resistance: A History of Land Tenure in New Mexico*. Norman: University of Oklahoma Press.

Ebright, M. 1994. *Land Grants and Lawsuits in Northern New Mexico*. Albuquerque: University of New Mexico Press.

Ebright, M., Hendricks, R., and Hughes, R. W. 2014. *Four Square Leagues: Pueblo Indian Land in New Mexico*. Albuquerque: University of New Mexico Press.

Echeverria, J. D. 2014. The Intersection of Water Law and Takings Doctrine. *Proceedings of the 60th Annual Rocky Mountain Mineral Law Institute*. 8B1–8B22. www-assets.vermontlaw.edu/Assets/directories/FacultyDocuments/Echeverria_IntersectionWaterLaw AndTakings.pdf.

Elliot, T. S. 1943. *Four Quartets*. New York: Harcourt.

Emel, J., and Brooks, E. 1988. Changes in Form and Function of Property Rights Institutions under Threatened Resource Scarcity. *Annals of the Association of American Geographers* 78(2): 241–252.

Emel, J., and Roberts, R. 1995. Institutional Form and Its Effect on Environmental Change: The Case of Groundwater in the Southern High Plains. *Annals of the Association of American Geographers* 85(4): 664–683

Emel, J., Roberts, R., and Saurí, D. 1992. Ideology, Property, and Groundwater Resources: An Exploration of Relations. *Political Geography* 11: 37–54.

Espeland, W. N. 1998. *The Struggle for Water: Politics, Rationality, and Identity in the American Southwest*. Chicago: University of Chicago Press.

Feller, J. M. 2007. The Adjudication That Ate Arizona Water Law. *Arizona Law Review* 49:405–440.

Ferguson, J., and Gupta, A. 2002. Spatializing States: Towards an Ethnography of Neoliberal Governmentality: American Ethnologist 29(4): 981–1002.

Fernald, A. G., Baker, T. T., and Guldan, S. J. 2007. Hydrologic, Riparian, and Agroecosystem Functions of Traditional Acequia Irrigation Systems. *Journal of Sustainable Agriculture* 30(2): 147–171.

Fernald, A. G., Tidwell, V., Rivera, J., Rodriguez, S., Guldan, S., Steele, C., Ochoa, C., Hurd, B., Ortiz, M., Boykin, K., and Cibils, A. 2012. Modeling Sustainability of Water, Environment, Livelihood, and Culture in Traditional Irrigation Communities and Their Linked Watersheds. *Sustainability* 4(11): 2998–3022.

Fleck, J. 2016. *Water Is for Fighting Over and Other Myths about Water in the West*. Washington, DC: Island Press.

Follansbee, R. 1919. *A History of the Water Resources Branch, U.S. Geological Survey; Volume I, from Predecessor Surveys to June 30, 1919*. Republished in 1994. Denver: United States Geological Survey.

Freyfogle, E. T. 2003. *The Land We Share: Private Property and the Common Good*. Washington, DC: Island Press.

Gleick, P. 2010. Roadmap for Sustainable Water Resources in Southwestern North America. *PNAS* 107(50): 21300–21305.

Glennon, R. 2002. *Water Follies: Groundwater Pumping and the Fate of America's Fresh Waters*. Washington, DC: Island Press.

———. 2006. General Stream Adjudications and the Environment. *Journal of Contemporary Water Research and Education* 133(1): 26–28.

Graf, W. 1994. *Plutonium and the Rio Grande: Environmental Change and Contamination in the Nuclear Age*. New York: Oxford University Press.

Grantham, T. E., and Viers, J. 2014. 100 Years of California's Water Rights System: Patterns, Trends, and Uncertainty. *Environmental Research Letters* 9:1–10.

Groenfeldt, D. 2010. The Next Nexus: Environmental Ethics, Water Management and Climate Change. *Water Alternatives* 3(3): 575–586.

Gupta, J., Dellapenna, J. W., and van den Heuvel, M. 2016. Water Sovereignty and Security, High Politics and Hard Power: The Dangers of Borrowing Discourses! In *Handbook on Water Security*, eds. C. Pahl-Wostl, A. Bhaduri, and J. Gupta, pp. 120–136. Cheltenham, UK: Edward Elgar.

Hall, G. E. 2002. *High and Dry: The Texas-New Mexico Struggle for the Pecos River*. Albuquerque: University of New Mexico Press.

Hannah, M. 2000. *Governmentality and the Mastery of Territory in Nineteenth-Century America*. Cambridge: Cambridge University Press.

Harden, B. 1997. *A River Lost: The Life and Death of the Columbia*. New York: W. W. Norton.

Harley, J. B. 1988. Maps, Knowledge, and Power. In *The Iconography of Landscape: Essays on the Symbolic Representation, Design and Use of Past Environments*, eds. Denis Cosgrove and Stephen Daniels, pp. 277–312. Cambridge Studies in Historical Geography 9. Cambridge: Cambridge University Press.

Harvey, D. 2003. *The New Imperialism*. Cambridge: Oxford University Press.

Hess, D. J., Wold, C. A., Hunter, E., Nay, J., Worland, S., Gilligan, J., and Gornberger, G. M. 2016. Drought, Risk, and Institutional Politics in the American Southwest. *Sociological Forum* 31(S1): 807–827.

Heynen, N., McCarthy, J., Prudham, S., Robbins, P., eds. 2007. *Neoliberal Environments: False Promises and Unintended Consequences*. New York: Routledge.

Hill, N. S., and Ratteree, K., eds. 2017. *The Great Vanishing Act: Blood Quantum and the Future of Native Nations*. Golden, CO: Fulcrum.

Hughes, R. W. 2017. Pueblo Indian Water Rights: Charting the Unknown. *Natural Resources Journal* 57(1): 219–261.

Hutchins, W. A. 1955. The New Mexico Law of Water Rights. *Technical Report 4*. Santa Fe: Office of the State Engineer.

Ingram, B. Lynn, and Malamud-Roam, F. 2013. *The West without Water: What Past Floods, Droughts, and Other Climatic Clues Tell Us about Tomorrow.* Berkeley: University of California Press.

Ingram, H. 1990. *Water Politics: Continuity and Change.* Albuquerque: University of New Mexico Press.

Ingram, H., and Brown, F. L. 1987. The Community Value of Water: Implications for the Rural Poor in the Southwest. *Journal of the Southwest* 29(2): 179–202.

IPCC (Intergovernmental Panel on Climate Change). 2013. *Climate Change 2013: The Physical Science Basis. Contribution of Working Group I to the Fifth Assessment Report of the Intergovernmental Panel on Climate Change,* eds. T. F. Stocker, D. Qin, G.-K. Plattner, M. Tignor, S. K. Allen, J. Boschung, A. Nauels, Y. Xia, V. Bex, and P. M. Midgley. Cambridge: Cambridge University Press.

Jepson, W. 2012. Claiming Space, Claiming Water: Contested Legal Geographies of Water in South Texas. *Annals of the Association of American Geographers* 102(3): 614–631.

Jones, J. 2016. *A Little Dam Problem.* Caldwell, ID: Caxton Press.

Jones, P. A., and Cech, T. 2009. *Colorado Water Law for Non-lawyers.* Boulder: University Press of Colorado.

Kauanui, J. K. 2008. *Hawaiian Blood: Colonialism and the Politics of Sovereignty and Indigeneity.* Durham, NC: Duke University Press.

Kosek, J. 2006. *Understories: The Political Life of Forests in Northern New Mexico.* Durham, NC: Duke University Press.

Krause, F., and Strang, V. 2016. Thinking Relationships through Water. *Society and Natural Resources* 29(6): 633–638.

Kupel, D. E. 2003. *Fuel for Growth: Water and Arizona's Urban Environment.* Tucson: University of Arizona Press.

Lane, M. 2011. Water, Technology, and the Courtroom: Negotiating Reclamation Policy in Territorial New Mexico. *Journal of Historical Geography* 37(3): 300–311.

Latour, B. 2010. *The Making of Law: An Ethnography of the Conseil d'état.* Malden, MA: Polity Press.

Lave, R. 2012. *Fields and Streams: Stream Restoration, Neoliberalism, and the Future of Environmental Science.* Athens: University of Georgia Press.

———. 2015. Introduction to Special Issue on Critical Physical Geography. *Progress in Physical Geography* 39(5): 571–575.

Levine, F. 1990. Dividing the Water: The Impact of Water Rights Adjudication on New Mexican Communities. *Journal of the Southwest* 32(3): 268–277.

———. 2008. Listening to the Land: Tradition and Change in the Northern Rio Grande. In *Survival along the Continental Divide,* ed. J. Loeffler, pp. 255–265. Albuquerque: University of New Mexico Press.

Limerick, P., and Hanson, J. L. 2012. *A Ditch in Time: The City, the West, and Water.* Golden, CO: Fulcrum Press.

Linton, J. 2008. Is the Hydrologic Cycle Sustainable? A Historical-Geographical Critique of a Modern Concept. *Annals of the Association of American Geographers* 98(3): 630–649.

———. 2010. *What Is Water? The History of a Modern Abstraction.* Vancouver: University of British Columbia Press.

Linton, J., and Budds, J. 2014. The Hydrosocial Cycle: Defining and Mobilizing a Relational-Dialectical Approach to Water. *Geoforum* 57:170–180.

Liverman, D. 2004. Who Governs, at What Scale and at What Price? Geography, Environmental Governance, and the Commodification of Nature. *Annals of the Association of American Geographers* 94(4): 734–738.

Livingstone, D. 2003. *Putting Science in Its Place: Geographies of Scientific Knowledge.* Chicago: University of Chicago Press.

MacDonnell, L. J. 2015. Rethinking the Use of General Stream Adjudications. *Wyoming Law Review* 15(2): 347–381.

MacKenzie, D. 2006. *An Engine, Not a Camera: How Financial Models Shape Markets.* Cambridge, MA: MIT Press.

Margolis, E. Q., Meko, D. M., and Touchan, R. 2011. A Tree-Ring Reconstruction of Streamflow in the Santa Fe River, New Mexico. *Journal of Hydrology* 397:118–127.

Masco, J. 2006. *The Nuclear Borderlands: The Manhattan Project in Post-Cold War New Mexico.* Princeton, NJ: Princeton University Press.

Mason, M., and Khawlie, M. 2016. Fluid Sovereignty: State-Nature Relations in the Hasbani Basin, Southern Lebanon. *Annals of the American Association of Geographers* 106(6): 1344–1359.

Mathews-Lamb, S. K. 2001. "Between This River and That": Establishing Water Rights in the Chama Basin of New Mexico. In *Fluid Arguments,* ed. C. Miller, pp. 40–61. Tucson: University of Arizona Press.

Mattei, U., and Nader, L. 2008. *Plunder: When the Rule of Law is Illegal.* Oxford: Blackwell.

Matthews, P. O. 1984. *Water Resources, Geography, and Law.* Washington, DC: Association of American Geographers.

———. 2003. Simplifying Western Water Rights to Facilitate Water Marketing. *Water Resources Update* 126(1): 40–44.

McCarthy, J. 2005. Commons as Counterhegemonic Projects. *Capitalism Nature Socialism* 16(1): 9–24.

McCool, D. 1987. *Command of the Waters: Iron Triangles, Federal Water Development, and Indian Water.* Tucson: University of Arizona Press.

———. 2002. *Native Waters: Contemporary Indian Water Settlements and the Second Treaty Era.* Tucson: University of Arizona Press.

———. 2012. *River Republic: The Fall and Rise of America's Rivers.* New York: Columbia University Press.

McGranahan, C. 2016. Theorizing Refusal: An Introduction. *Cultural Anthropology* 31(3): 319–325.

McKinney, M., and Thorson, J. E. 2015. Resolving Water Conflicts in the American West. *Water Policy* 17:679–706.

McPhee, J. 1989. *The Control of Nature.* New York City: Farrar, Strauss, and Giroux.

Meehan, K. 2014. Tool-Power: Water Infrastructure as Wellsprings of State Power. *Geoforum* 57:215–224.

Meehan, K., and Moore, A. W. 2014. Downspout Politics, Upstream Conflict: Formalizing Rainwater Harvesting in the United States. *Water International* 39:417–430.

Meinzen-Dick, R. 2007. Beyond Panaceas in Water Institutions. *PNAS* 104(39): 15200–15205.

Meizen-Dick, R., and Mwangi, E. 2008. Cutting the Web of Interests: Pitfalls of Formalizing Property Rights. *Land Use Policy* 26:36–43.

Merchant, C. 2016. *Autonomous Nature: Problems of Prediction and Control from Ancient Times to the Scientific Revolution.* New York: Routledge Press.

Merrill, T. W. 2011. The Property Prism. *Econ Journal Watch* 8(3): 247–254.

Meyer, M. C. 1984. *Water in the Hispanic Southwest: A Social and Legal History, 1550–1850.* Tucson: University of Arizona Press.

Miller, C. Y. 2003. Death Knell Rings for the PIA?: In Re General Adjudication of All Right to Use Water in Gila River System and Source. *Arizona Law Review* 45:241–245.

Mitchell, T. 2002. *The Rule of Experts: Egypt, Techno-Politics, Modernity.* Berkeley: University of California Press.

Molle, F. 2009. River-Basin Planning and Management: The Social Life of a Concept. *Geoforum* 40(3): 484–494.

Montoya, M. 2002. *Translating Property: The Maxwell Land Grant and the Conflict over Land in the American West, 1840–1900.* Lawrence: University Press of Kansas.

Moore, L. 2013. *Common Ground on Hostile Turf: Stories from an Environmental Mediator.* Washington, DC: Island Press.

Mulvaney, M. E. 2005. State Ex Rel. Martinez v. City of Las Vegas: The Misuse of History and Precedent in the Abolition of the Pueblo Water Rights Doctrine in New Mexico. *Natural Resources Journal* 45:1089–1116.

Murphy, A. (2013). Territory's Continuing Allure. *Annals of the Association of American Geographers* 103(4): 1212–1226.

New Mexico Office of the State Engineer. 1968–1969. *Taos Hydrographic Survey.* Santa Fe: New Mexico Office of the State Engineer. Digital files and maps available at www.ose .state.nm.us/HydroSurvey/legal_ose_hydro_Taos.php.

———. 2003. *New Mexico State Water Plan: Working Together towards Our Water Future.* Santa Fe: New Mexico Office of the State Engineer. www.ose.state.nm.us/Planning /SWP/PDF/2003StateWaterPlan.pdf.

———. 2012. Settlement Agreement (Aamodt). Document available at the NM OSE website. www.ose.state.nm.us/Legal/settlements/Aamodt/aamodt_agreement.php. Accessed January 15, 2018.

———. 2012. Taos Pueblo Water Rights Settlement Agreement. Document available at www.ose.state.nm.us/Legal/settlements/Taos/Taos_documents.php. Accessed January 16, 2018.

———. 2013. New Mexico Water Use by Categories 2010. *Technical Report 54.* Santa Fe: New Mexico Office of the State Engineer.

Nichols, J. 2012. Aamodt, Schmaamodt: Who Really Gets the Water? In *One Hundred Years of Water Wars in New Mexico, 1912–2012,* ed. C. T. Ortega Klett, pp. 166–179. Santa Fe: Sunstone Press.

Nieto-Phillips, J. M. 2008. *The Language of Blood: The Making of Spanish-American Identity in New Mexico, 1850–1940.* Albuquerque: University of New Mexico Press.

NMSA (New Mexico Statutes Annotated). 2008. Michie's Annotated Statutes of New Mexico, Chapter 72 and Chapter 73. Issued by the New Mexico Acequia Association. Charlottesville, VA: LexisNexis.

Norman, E., and Bakker, K. 2009. Transgressing Scales: Transboundary Water Governance across the Canada-U.S. border. *Annals of the Association of American Geographers* 99(1): 99–117.

Nostrand, R. L. 1992. *The Hispano Homeland*. Norman: University of Oklahoma Press.

Offen, K. 2004. Historical Political Ecology: An Introduction. *Historical Geography* 32:19–42.

O'Neill, K. M. 2006. *Rivers by Design: State Power and the Origins of U.S. Flood Control*. Durham, NC: Duke University Press.

Ortega Klett, C. T., ed. 2012. *One Hundred Years of Water Wars in New Mexico, 1912–2012*. Santa Fe: Sunstone Press.

Ortiz, M., Brown, C., Fernald, A., Baker, T. T., Creel, B., and Guldan, S. 2007. Land Use Change Impacts on Acequia Water Resources in Northern New Mexico. *Journal of Contemporary Water Research and Education* 137(1): 47–54.

Parenti, C. 2015. The Environment Making State: Territory, Nature, and Value. *Antipode* (47): 829–848.

Pease, M. 2010. Constraints to Water Transfers in Unadjudicated Basins: The Middle Rio Grande as a Case Study. *Journal of Contemporary Water Research and Education* 144(1): 37–43.

Pecos, R. 2007. The History of Cochiti Lake from the Pueblo Perspective. *Natural Resources Journal* 47:639–652.

Peña, D., ed. 1999. *Chicano Culture, Ecology, Politics: Subversive Kin*. Tucson: University of Arizona Press.

Perramond, E. 2012. The Politics of Scaling Water Governance and Adjudication in New Mexico. *Water Alternatives* 5(1): 62–82.

———. 2013. Water Governance in New Mexico: Adjudication, Law, and Geography. *Geoforum* 45:83–93.

Perramond, E., and Lane, M. (2014). Territory to State: Law, Power, and Water in New Mexico. In *Negotiating Territoriality: Spatial Dialogues between State and Tradition*, eds. A. Dawson, L. Zanotti, and I. Vaccaro, pp. 142–159. New York: Routledge Press.

Perreault, T. 2005. State Restructuring and the Scale Politics of Rural Water Governance in Bolivia. *Environment and Planning A* 37(2): 263–284.

———. 2008. Custom and Contradiction: Rural Water Governance and the Politics of "Usos Y Costumbres" in Bolivia's Irrigators' Movement. *Annals of the Association of American Geographers* 98(4): 835–854.

Phillips, F. M., Hall, G. E., Black, M. E. 2011. *Reining in the Rio Grande: People, Land, and Water*. Albuquerque: University of New Mexico Press.

Pisani, D. 2002. *Water and American Government: The Reclamation Bureau, National Water Policy, and the West, 1902–1935*. Berkeley: University of California Press.

Plewa, T. M. 2009. *A Trickle Runs through It: An Environmental History of the Santa Fe River, New Mexico*. PhD diss., University of South Carolina, Columbia.

Price, V. B. 2011. *The Orphaned Land: New Mexico's Environment Since the Manhattan Project*. Albuquerque: University of New Mexico Press.

Prieto, M. 2016. Practice *Costumbres* and the Decommodification of Nature: The Chilean Water Markets and the Atacameño People. *Geoforum* 77:28–39.

Quintana, F. L. 1990. Land, Water, and Pueblo-Hispanic Relations in Northern New Mexico. *Journal of the Southwest* 32(3): 288–299.

Reich, P. L. 1994. Mission Revival Jurisprudence: State Courts and Hispanic Water Law Since 1850. *Washington Law Review* 69(4): 869–925.

Reno-Trujillo, M., Tidwell, V., Broadben, C., Brookshire, D., Coursey, D., Jackson, C., Polley, A., and Stevenson, B. 2013. *Designing a Water Leasing Market for the Mimbres River, New Mexico.* SAND2013-3556. Albuquerque: Sandia National Laboratories.

Reisner, M. 1986. *Cadillac Desert: The American West and Its Disappearing Water.* New York: Viking Press.

Reuss, M. 2008. Seeing Like an Engineer. *Technology and Culture* 49:531–546.

Richards, E. 2005. Water Rights Settlement Agreements in New Mexico: What Are They, Why Have They Emerged, How Have They Been Achieved, and What Do They Imply for the Future? *New Mexico Water Dialogue Newsletter,* 2–9.

Ridgeley, G. C. 2010. *The Future of Water Adjudications in New Mexico.* Paper delivered at the 55th Annual NM Water Conference, How Will Institutions Evolve to Meet Our Water Needs in the Next Decade? Las Cruces.

Rivera, J. 1998. *Acequia Culture.* Albuquerque: University of New Mexico Press.

Robbins, P. 2000. The Practical Politics of Knowing: State Environmental Knowledge and Local Political Economy. *Economic Geography* 76(2): 126–144.

Robison, J., Cosens, B., Jackson, S., Leonard, K., and McCool, D. 2018, forthcoming. Indigenous Water Justice. *Lewis and Clark Law Review* 42.

Robertson, M. 2010. Performing Environmental Governance. *Geoforum* 41(1): 7–10.

Rockström, J., Falkenmark, M., Allan, T., Folke, C., Gordon, L., Jägerskog, A., Kummu, M., Lannerstad, M., Meybeck, M., Molden, D., Postel, S., Savenije, H. H. G., Svedin, U., Turton, A., and Varis, O. 2014. The Unfolding Water Drama in the Anthropocene: Towards a Resilience-Based Perspective on Water for Global Sustainability. *Ecohydrology* 7:1249–1261.

Rodriguez, S. 1990. Applied Research on Land and Water in New Mexico: A Critique. *Journal of the Southwest* 32(3): 300–315.

———. 2006. *Acequia: Water-Sharing, Sanctity and Place.* Santa Fe: School for Advanced Research Press.

Romero Pike, M. 2010. Memorias del Río. In *The Return of the Rivers: Writers, Scholars, and Citizens Speak on Behalf of the Santa Fe River,* ed. A. Kyce Bello, pp. 39–44. Santa Fe: Sunstone Press.

Romero-Wirth, C., and Kelly, S. 2012. *Water Rights Management in New Mexico and along the Rio Grande: Is AWRM Sufficient?* Albuquerque: Utton Transboundary Resources Center, University of New Mexico.

Roth, D., Boelens, R., and Zwarteveen, M., eds. 2005. *Liquid Relations: Contested Water Rights and Legal Complexity.* New Brunswick, NJ: Rutgers University Press.

Roybal, K. 2017. *Archives of Dispossession: Recovering the Testimonios of Mexican American Herederas, 1848–1960.* Chapel Hill: University of North Carolina Press.

Samani, Z., Skaggs, R., and Longworth, J. 2013. Alfalfa Water Use and Crop Coefficients across the Watershed: From Theory to Practice. *Journal of Irrigation and Drainage Engineering* 139(5): 341–348.

Sangameswaran, P. 2009. Neoliberalism and Water Reforms in Western India: Commercialization, Self-Sufficiency, and Regulatory Bodies. *Geoforum* 40(2): 228–238.

Saranillio, D. I. 2015. Settler Colonialism. In *Native Studies Keywords,* eds. S. N. Teves, A. Smith, and M. Raheja, pp. 284–300. Tucson: University of Arizona Press.

Saurí, D. 1990. *From Mayordomos to State Engineers: Historical Change in New Mexico Water Rights.* Ph.D. diss., Graduate School of Geography, Clark University, Worcester, MA.

Schmidt, J. J. 2017. *Water: Abundance, Scarcity, and Security in the Age of Humanity.* New York: New York University Press.

Schmidt, J. J., and Mitchell, K. R. 2014. Property and the Right to Water: Toward a Nonliberal Commons. *Review of Radical Political Economics* 46(1): 54–69.

Schorr, D. 2012. *The Colorado Doctrine: Water Rights, Corporations, and Distributive Justice on the American Frontier.* New Haven, CT: Yale University Press.

Scott, J. 1998. *Seeing Like a State: How Certain Schemes to Improve the Human Condition Have Failed.* New Haven, CT: Yale University Press.

———. 2009. *The Art of Not Being Governed: An Anarchist History of Upland Southeast Asia.* New Haven, CT: Yale University Press.

Sedlak, D. 2014. *Water 4.0: The Past, Present, and Future of The World's Most Vital Resource.* New Haven, CT: Yale University Press.

Shiveley, D. 2001. Water Right Reallocation in New Mexico's Rio Grande Basin, 1975–1995. *Water Resources Development* 17(3): 445–460.

Simmons, M. 1972. Spanish Irrigation Practices in New Mexico. *New Mexico Historical Review* 47(2): 135–150.

Simon, G. 2017. *Flame and Fortune in the American West: Urban Development, Environmental Change, and the Great Oakland Hills Fire.* Berkeley: University of California Press.

Skaggs, R., Samani, Z., Salim Bawazir, A., and Bleiweiss, M. 2011. The Convergence of Water Rights, Structural Change, Technology, and Hydrology. *Natural Resources Journal* 51:95–117.

Snow, D. 1988. *The Santa Fe Acequia Systems.* Santa Fe: City of Santa Fe Planning Department.

Strang, V. 2009. *Gardening the World: Agency, Identity, and the Ownership of Water.* New York: Bergahn Books.

Stroup, L. J. 2011. Adaptation of U.S. Water Management to Climate and Environmental Change. *Professional Geographer* 63(4): 414–428.

Sugg, Z., Ziaja, S., and Schlager, E. C. 2016. Conjunctive Groundwater Management as a Response to Socio-ecological Disturbances: A Comparison of 4 Western U.S. States. *Texas Water Journal* 7(1): 1–24.

Swyngedouw, E. 2004. *Social Power and the Urbanization of Water: Flows of Power.* Oxford: Oxford University Press.

———. 2005. Dispossessing H2O: The Contested Terrain of Water Privatization. *Capitalism Nature Socialism* 16(1): 81–98.

———. 2009. The Political Economy and Political Ecology of the Hydrosocial Cycle. *Journal of Contemporary Water Research and Education* 142:56–60.

TallBear, K. 2013. *Native American DNA: Tribal Belonging and the False Promise of Genetic Science.* Minneapolis: University of Minnesota Press.

Tarlock, D. 1985. The Endangered Species Act and Western Water Rights. *Land and Water Law Review* 20(1): 1–30.

———. 1989. The Illusion of Finality in General Stream Adjudications. *Idaho Law Review* 25:271–289.

———. 2006. General Stream Adjudications: A Good Public Investment? *Journal of Contemporary Water Research and Education* 1333:52–61.

———. 2010. How Well Can Water Law Adapt to the Potential Stresses of Global Climate Change? *University of Denver Water Law Review* 14:1.

Taylor P. L., and Sonnenfeld, D. A. 2017. Water Crises and Institutions: Inventing and Reinventing Governance in an Era of Uncertainty. *Society and Natural Resources* 30(4): 395–403.

Thorson, J. E. 2001. Clarifying State Water Rights and Adjudications. *Two Decades of Water Law and Policy Reform: A Retrospective and Agenda for the Future (Summer Conference, June 13–15)*. Boulder: University of Colorado.

Thorson, J. E., Kropf, R., Gerlak, A. K., and Crammond, D. 2005. Dividing Western Waters: A Century of Adjudicating Rivers and Streams, Part I. *University of Denver Water Law Review* 8(2): 355–461.

———. 2006. Dividing Western Waters: A Century of Adjudicating Rivers and Streams, Part II. *University of Denver Water Law Review* 9(2): 299–484.

Tidwell, V. C., Moreland, B. D., Zemlick, K. M., Roberts, B. L., Passell, H. D., Jensen, D., Forsgren, C., Sehlke, G., Cook, M. A., King, C. W., and Larsen, S. 2014. Mapping Water Availability, Projected Use and Cost in the Western United States. *Environmental Research Letters* 9:1–11.

Titus, F. 2005. On Regulating New Mexico's Domestic Wells. *Natural Resources Journal* 45: 853–863.

Turner, M. 2014. Political Ecology I: An Alliance with Resilience? *Progress in Human Geography* 38(4): 616–623.

———. 2016. Political Ecology III: The Commons and Communing. *Progress in Human Geography* 41(6): 1–8.

Tyler, Daniel. 1990. *The Mythical Pueblo Rights Doctrine: Water Administration in Hispanic New Mexico*. El Paso, TX: Texas Western Press.

Utton Center. 2013. *Water Matters!* Albuquerque: University of New Mexico Utton Law School.

Valentine, J. 2012. Adjudications: Managing Water Wars in New Mexico. In *One Hundred Years of Water Wars in New Mexico, 1912–2012*, ed. C. T. Ortega Klett, pp. 29–51. Santa Fe: Sunstone Press.

Veracini, L. 2010. *Settler Colonialism: A Theoretical Overview*. Hampshire, UK: Palgrave MacMillan.

Vlasich, J. A. 2005. *Pueblo Indian Agriculture*. Albuquerque: University of New Mexico.

Wainwright, J., and Robertson, M. 2003. Territorialization, Science and the Colonial State: The Case of Highway 55 in Minnesota. *Cultural Geographies* 10:196–217.

Ward, E. R. 2015. *Border Oasis: Water and the Political Ecology of the Colorado River Delta, 1940–1975*. Tucson: University of Arizona Press.

Wescoat, J. L. Jr. 1995. The Right of Thirst for Animals in Islamic Law: A Comparative Approach. *Environment and Planning D: Society and Space* 13(6): 637–654.

Wescoat, J. L. Jr., Headington, L., and Theobald, R. 2007. Water and Poverty in the United States. *Geoforum* 38(5): 801–814.

Western States Water Council. 2011. Indian Water Rights Settlements approved by Congress. www.westernstateswater.org/publications/. Accessed January 15, 2018.

White, R. 1995. *The Organic Machine: The Remaking of the Columbia River.* New York: Hill and Wang.

Whitehead, M., Jones, M., and Jones R., 2007. *The Nature of the State: Excavating the Political Ecologies of the Modern State.* Oxford: Oxford University Press.

Whiteley, J., Ingram, H., and Perry, R., eds. 2008. *Water, Place, and Equity.* Cambridge, MA: MIT Press.

Wilder, M., and Lankao, P. R. 2006. Paradoxes of Decentralization: Neoliberal Reforms and Water Institutions in Mexico. *World Development* 34(11): 1977–1995.

Wilder, M., Scott, C., Pablos, N. P., Varady, R., Garfin, G., and McEvoy, J. 2010. Adapting across Boundaries: Climate Change, Social Learning, and Resilience in the US-Mexico Border Region. *Annals of the Association of American Geographers* 100(4): 917–928.

Wilmsen, C. 2007. Maintaining the Environmental-Racial Order in Northern New Mexico. *Environment and Planning D: Society and Space* 25(2): 236–257.

Wolf, A. 2000. Indigenous Approaches to Water Conflict Negotiations and Implications for International Waters. *International Negotiation* 5(2): 357–373.

Wolf, E. 1982. *Europe and the People without History.* Berkeley: University of California Press.

Wolfe, P. 2006. Settler Colonialism and the Elimination of the Native. *Journal of Genocide Research* 8(4): 387–409.

Woodhouse, C. A., Meko, D. M., Griffin, D., and Castro, C. L. 2013. Tree Rings and Multiseason Drought Variability in the Lower Rio Grande Basin, USA. *Water Resources Research* 49(2): 844–850.

Worster, D. 1985. *Rivers of Empire: Water, Aridity, and the Growth of the American West.* New York: Pantheon Books.

———. 2002. *A River Running West: The Life of John Wesley Powell.* Oxford: Oxford University Press.

———. 2003. Watershed Democracy: Recovering the Lost Vision of John Wesley Powell. *Journal of Land, Resources, and Environmental Law* 23(1): 57–66.

Zetland, D. 2011. *The End of Abundance: Economic Solutions to Water Scarcity.* n.p.: Aguanomics Press.

———. 2014. *Living with Water Scarcity.* n.p.: Aguanomics Press.

INDEX

Figures, tables, and maps are indicated by page numbers followed by *fig., tab., map.* Notes are indicated by page numbers followed by n.

practicably irrigable acreage (PIA), 23, 38–39
Price, V. B., 83, 196n25
Prieto, M., 211n38
prior appropriation law: archival sources and,
    118; beneficial use of water and, 25–26; lack of
    incentives for water conservation in, 25–26,
    112; OSE enforcement of, 96; water code
    (1907), 22–23; water expertise and, 92
private adjudicatory industries: consulting histo-
    rians and, 121; general stream adjudications
    and, 110, 200n8; lawyers and, 121; public
    monies and, 120–21, 200n8; settlements
    and, 181
property: common-use resistance to, 203n25;
    community-based, 19; field mapping and,
    113; gendered dispossession of, 15, 189n6; per-
    formative aspects of, 26, 202n11; state and, 47,
    178; water as, 4, 6, 16, 34, 89, 102, 104–5, 116,
    125, 174, 177–79
proportional allocation, 184
Prudham, S., 190n14
Pueblo Indians: acequias and, 15; federal legal
    counsel and, 40, 52–53, 68, 77–78; federally
    protected sovereignty, 29–30, 37; floodwater
    farming and, 36; historical beneficial use
    rights, 38–39, 53; as Mexican citizens, 27–28,
    38–39; prior appropriation law and, 153;
    quantification of water rights, 36, 38, 40, 126,
    145, 152, 157, 178; redefinition as Indians, 27,
    38; relations with state and, 37–38; resource
    justice and, 81; Rio Grande River and, 156–57;
    senior water rights and, 23, 37; Spanish set-
    tlers and, 27; stewardship of, 157–58; water
    cultures of, 202n3; water governance and,
    84–85; water justice and, 36, 53, 81, 161, 178,
    183; water quality and, 182–83; water rights
    claims, 37–39, 44; water sovereignty and,
    187n2
Pueblo Lands Board, 38
Pueblo Revolt (1680), 27
pueblo rights doctrine: city use of, 151, 204n3,
    204n5, 204n7; myth of, 146; Spanish settlers
    and, 119, 146
PVWUA. See Pojoaque Valley Water Users As-
    sociation (PVWUA)

querencia (sense of place), 36

Raise the River, 172
Ratteree, K., 191n36
Reagan, Ronald, 40
Reisner, M., 188n9

regional acequia associations: adjudication and,
    59, 130; civil society groups and, 129–130,
    132–34; legal aid for, 133–34; location of,
    131map; mutual domestic water associations
    and, 132; regulations and, 130, 132; response
    to state by, 142; settlement and, 65, 69; types
    of, 130; water shortages and, 98
regional water planning: added infrastructure
    and, 140; civil society groups and, 140;
    development of, 136–37; distrust of, 139–140;
    Interstate Stream Commission (ISC) and,
    137, 139; jurisdictional fragmentation in,
    137–39; New Mexico, 136–39; process of,
    140; stakeholder participation and, 137, 139;
    state water plan and, 139; water regions in,
    138map; watersheds and, 138; well develop-
    ment restrictions and, 51
Reno-Trujillo, M, 198n27
resource justice, 81
Reuss, Martin, 194n7, 196n27
Reynolds, Steve, 34, 50
Ridgeley, G. C., 200n4
Rio Arriba County, 141
Rio Embudo, 173, 183
Rio en Medio: conflict and, 47–48, 50; convenios
    for, 45–47, 47fig., 48, 49fig.; prior appro-
    priation law and, 45–48, 50; splitter box for,
    45–46; water sharing in, 45–47, 47fig., 48
Rio Gallinas Basin: special masters and, 98–100;
    water expertise and, 97; water measurements
    and, 98–99; water rights adjudications in, 98,
    100, 105
Rio Grande Basin, 34, 57, 73
Rio Grande Compact, 74, 76, 91–92, 169
Rio Grande Pueblo, 145
Rio Grande River: Chama project water and,
    71; climate change and, 163; flow in, 181;
    groundwater equivalency and, 73–74; Mexico
    and, 182; Nambé-Pojoaque-Tesuque (NPT)
    stream system and, 33; Otowi Gage and, 74,
    76; plutonium levels in, 166, 207n18; Pueblo
    Indians and, 27, 39, 152–53, 156–57; regional
    water system and, 51, 74, 77; return human
    flows in, 168–69; San Juan-Chama Project
    and, 67; water flow measurement, 90; water
    marketing and, 104; water rights and, 155
Rio Mimbres: individual rights and, 177;
    measurement and, 101–2; settlement in, 100;
    special masters and, 100, 102; water expertise
    and, 97; water markets and, 100, 102–3, 105;
    water rights adjudications in, 25–26, 100;
    water shortages in, 101

www.ingramcontent.com/pod-product-compliance
Lightning Source LLC
Chambersburg PA
CBHW030400270326
41926CB00009B/1201